African American Women Chemists in the Modern Era

African American Women Chemists in the Modern Era

JEANNETTE E. BROWN

OXFORD
UNIVERSITY PRESS

OXFORD
UNIVERSITY PRESS

Oxford University Press is a department of the University of Oxford. It furthers
the University's objective of excellence in research, scholarship, and education
by publishing worldwide. Oxford is a registered trade mark of Oxford University
Press in the UK and certain other countries.

Published in the United States of America by Oxford University Press
198 Madison Avenue, New York, NY 10016, United States of America.

© Oxford University Press 2018

Library of Congress Cataloging-in-Publication Data
Names: Brown, Jeannette E. (Jeannette Elizabeth), 1934– author.
Title: African American women chemists in the modern era / Jeannette E. Brown.
Description: New York, NY : Oxford University Press, [2018] |
Includes bibliographical references and index.
Identifiers: LCCN 2018018332 | ISBN 9780190615178
Subjects: LCSH: African American women chemists—Biography. |
Chemists—United States—Biography.
Classification: LCC QD21 .B6925 2018 | DDC 540.92/520973—dc23
LC record available at https://lccn.loc.gov/2018018332

9 8 7 6 5 4 3 2 1

Printed by Sheridan Books, Inc., United States of America

This book is dedicated to all the young girls and boys who are exploring the field of chemistry for a future occupation or for any reason. I hope you succeed in your chosen profession.

Contents

Foreword

IN 2012, I received an invitation to write a review of Jeannette E. Brown's book *African American Women Chemists* from the editor of the *Bulletin for the History of Chemistry*. As a chemist, I was both honored and humbled to write the review. This was the first book that provided an important perspective on the contributions of African American women in the chemical sciences. This matters because there are limited books and resources focused on the contributions of women of color in the STEM (Science, Technology, Engineering, Mathematics, and Engineering) disciplines.

The Double Bind: The Price of Being a Minority Woman in Science (1976) authored by Shirley Malcom, Paula Quick Ball, and Janet Walsh established a critical voice and platform for discussing the challenges faced by women of color in STEM. The book *African American Women Chemists* provided yet another voice to the ongoing national conversation of diversity and inclusion issues within the scientific community. Specifically, the book provided important biographies of chemists, who are often overlooked or ignored in the realm of history of science.

Historically, the intellectual contributions in STEM has focused on men, not women. In the African American community and mainstream media, pioneers such as George Washington Carver and Benjamin Banneker are frequently celebrated for their contributions while we continue to ignore the contributions of African American women. In recent years, we have begun to recognize the significant contributions of Dr. Marie Maynard Daly, the first African American woman to earn a PhD in chemistry, and Alice Augusta Ball, who discovered a treatment for leprosy (Hansen's disease) in the early 1900s. It is a wonderful historical fact that Daly is the *first*, but that is not why she matters. She is important because of her scientific contributions just like anybody else. Daly investigated the relationship between heart attacks and cholesterol in the 1950s. Today, the pharmaceutical industry generates significant revenue from drugs to treat cardiovascular disease, which emphasizes the importance of Daly's research on our society. My point is that Daly's contributions are just as important and significant as Carver's efforts in agricultural chemistry.

Brown's new book, *African American Women Chemists in the Modern Era,* Volume II, continues to enhance this important conversation by including contemporary chemists in academia, government, and industry. This new effort is certainly "beyond" Daly and Ball. The amazing chemists described in this new book have made important intellectual contributions in chemical education, science education and pedagogy, patent law and inventions, nanoparticle sensors, applications of photochemistry and spectroscopy, cancer research, medical chemistry, and marketing and new business development! Furthermore, these women have published in high impact journals such as the *Journal of the American Chemical Society, Journal of Chemical Education, Analytical Chemistry, Journal of Organic Chemistry, Inorganic Chemistry*, and the *Journal of Physical Chemistry B*, and have received patents.

I honestly believe that the historical biographies described in *African American Women Chemists in the Modern Era* can be viewed as a tool and resource for a successful career pathway in the chemical sciences, for all women. There are multiple pathways to success and everyone's journey is unique. Former Secretary of State Madeline Albright said it best when she stated, "There's a special place in hell for women who don't help each other!" Brown's book is a tool and yet another example for women of color in STEM to help each other.

Sibrina Collins, PhD
Executive Director
The Marburger STEM Center
Lawrence Technological University
21000 West Ten Mile Road
Southfield, Michigan 48075

Acknowledgments

BECAUSE THIS IS my second book on the subject, you would think it would have been easier to produce. In some respects it was, because I knew the job that I was facing, and because I was sent a booklet about the process of publication.

This book was written using the oral histories of living African American women chemists. I started taking the oral histories for this book about four years ago by asking the women that I knew if I could interview them for the book. I later met or was referred to other women that I could interview.

I want to thank all the women whose oral histories I collected, by interviewing them and recording their stories, all of which will be archived by the Science History Institute (SHI). I interviewed many more women than are included in the book, but their stories will still be archived by SHI. I want to say thank you to the women who took almost three hours of their busy lives to sit and talk to me about their work and personal lives. I enjoyed it and I hope you did too. I also want to thank these women for taking time to edit their stories, several times, and for the most part complying with my deadlines. I know I should not have been working on the book when many of you were on vacation.

The members of the SHI. Oral History Department transcribed all the oral histories, some at short notice, as I was finishing the book. I want to thank Center for Oral History of SHI, Director David Caruso and Curator Lee Berry and the rest of their staff for doing the tedious job of transcribing the three-hour-long oral histories and contacting the women to check the accounts for accuracy. I hope they will all allow the full story of their lives to be archived at SHI.

I would like to thank Olinda Young who was given the task of editing the story of each of the women in the book to make each of them readable and understandable to a lay person such as herself.

I would also like to thank the members of the American Chemical Society (ACS) Women Chemists Committee and Committee on Minority Affairs for their support and encouragement to write a second book. I also want to thank the members of the North Jersey ACS section for putting up with me as I focused on the book and not on my ACS work.

I want to thank the people who read and commented on my proposal for this book—twice. Most of them had read my first book and commented on the things that I should improve upon in the second book. I hope I have complied with their wishes.

I also want to thank Dr. Henry Lewis Gates Jr. who I consider a mentor and whom I was finally able to meet in person. Happily, he has agreed to write a review of this book for me.

Finally, I want to thank my editor Jeremy Lewis at Oxford University Press for being supportive of this second book and sending the first proposal back to me for editing. I also want to thank him for waiting while I worked to get my written accounts reviewed by the women despite the deadline. I hope the wait was worth it.

Finally, the women who have agreed to be in this book only scratch the surface. There are so many more women, some of whom I interviewed and some who have been interviewed by other people, whose stories could have been included in this book. However, since this is the digital age, their stories are available on the web and can be found with a little bit of searching.

I hope you enjoy reading this book.

Jeannette E. Brown
Hillsborough, NJ
July 27, 2017

1

Introduction

WHEN I WROTE my first book *African American Women Chemists* I neglected to state that it was a historical book. I researched to find the first African American woman who had studied chemistry in college and worked in the field. The woman that I found was Josephine Silane Yates who studied chemistry at the Rhode Island Normal School in order to become a science teacher. She was hired by the Lincoln Institute[1] in 1881 and later was, I believe, the first African American woman to become a professor and head a department of science. But then again there might be women who traveled out of the country to study because of racial prejudice in this country. The book ended with some women like myself who were hired as chemists in the industry before the Civil Rights Act of 1964.[2]

Therefore, I decided to write another book about the current African American women chemists who, as I say, are hiding in plain sight. To do this, I again researched women by using the web or by asking questions of people I met at American Chemical Society ACS or National Organization for the Professional Advances of Black Chemists and Chemical Engineers (NOBCChE) meetings. I asked women to tell me their life stories and allow me to take their oral history, which I recorded and which were transcribed thanks to the people at the Chemical Heritage Foundation[3] in Philadelphia, PA. Most of the stories of these women will be archived at the CHF in their oral history collection.

The women who were chosen to be in this book are an amazing group of women. Most of them are in academia because it is easy to get in touch with professors since they publish their research on the web. Some have worked for the government in the national laboratories and a few have worked in industry.

Some of these women grew up in the Jim Crow south where they went to segregated schools but were lucky because they were smart and had teachers and parents who wanted them to succeed despite everything they had to go through. Some even had problems in the northern schools and towns even though they might not have experienced de facto segregation.

Most managed to make it through college without problems. They had mentors and other professors who cared for them. One of the women made a career helping minority students succeed in college by teaching them how to study. Some managed to find other minority students in their class whom they could work with while in college. They also managed to be able to succeed well enough in college so that they could apply to graduate school, and most even have some post-graduate education.

Having said that, this book is about women who have made it in their careers and are beginning to give back to the minority community. I still wonder about those women who were not encouraged to try to become scientists or to study hard; the women who were told that they could only go to college to find a man to marry and take care of them; the women who did not even go to college and are struggling to make ends meet with any job because maybe their teachers thought they would never succeed.

All the women in this book that I interviewed told me their thoughts on what they would do now to give back to the community and help other minority students succeed. You will find this in the last paragraph of each of their stories. If you don't understand the chemistry they did, please skip to the last few pages of their story to see what they are doing to help students and young people to study STEM.[4]

Information about minority women in STEM

In 2010 the magazine *Diverse* published an article entitled "Survey: US Women and Minority Scientists Often Discouraged from Pursuing STEM Careers"[5] about the Bayer Facts of Science Education Survey, which investigated the root causes of underrepresentation of minorities in STEM fields by looking at the experiences of female, African American, Hispanic, and American Indian scientists from childhood through the workplace. The analysis reveals that young girls from minority groups have the same interest in science that all kids have, but they are not encouraged to study science. The stereotyping of those young women and the fact that they are mostly poorly educated in inferior schools are two reasons that STEM fields are not diverse. These girls did not see many scientists and engineers in their communities. Even if the girls and women were able to succeed in science, they faced a lack of self-confidence and encouragement from professional peers. As a result, in 2010, "More than three-quarters (77 percent) said significant numbers of women and under-represented minorities are missing from the US STEM workforce today because they were not identified, encouraged or nurtured to pursue STEM studies as youngsters."[6] Even if they were hired by a corporation, they could not advance for many reasons, including managerial bias and lack of advancement opportunities. Most of the women surveyed gave their corporation a C with respect to finding women or minorities in senior positions who could act as mentors to them. To read the survey summary go to Bayer Facts of Science Pressroom.[7]

Looking at recent data, I found an article "STEM Workforce No More Diverse Than 14 Years Ago" written by US News[8] with data from Change the Equation. People in the workplace are ageing, but young people are not in a position to replace them. One of the reasons for the absence of women and minorities may be the lack of Advanced Placement (AP) Courses or that students do not have the confidence to take these courses. Someone said this is squandered talent. For more information read the article.

There is another article in US News about women in STEM: "Report: No 'Leaky Pipeline' for Women in STEM,"[9] which says that even though women are underrepresented in STEM the fact that they leave after gaining a bachelor of science degree in college and do not pursue a PhD degree, may not be true, they may do other things with chemistry instead of working in industry or academia. Yes, they are underrepresented but some continue to pursue the degree.

Notes

1. Now Lincoln University in Missouri.
2. National Parks Service "Experience Your America" "Civil Rights Act of 1964." https://www.nps.gov/subjects/civilrights/1964-civil-rights-act.htm accessed May 9, 2018.
3. Chemical Heritage Foundation is now The Science History Institute.

4. Bayer Facts of Science Education Online Pressroom. http://bayerfactsofscience. online-pressroom.com/ accessed May 9, 2018.

5. "Survey: U. S. Women and Minority Scientis Often Dicouraged from Pursuing STEM Careers" March 23, 2010. diverseeducation.com/article/13644 accessed May 9, 2018.

6. Ibid.

7. Bayer Facts of Science Education Online Pressroom. http://bayerfactsofscience. onlinepressroom.com accesed May 9, 2018.

8. "STEM Workfoce No More Diverse than 14 Years Ago" by Allie Bidwell U.S. News. https://www.usnews.com/news/stem-solutions/articles/2015/02/24/stem-workforce-no-more-diverse-than-14-years-ago accessed May 9, 2018.

9. "No Leaky Pipeling for Women in STEM" by Allie Bidwell U. S. News. https://www.usnews.com/news/stem-solutions/articles/2015/02/17/report-no-leaky-pipeline-for-women-in-stem accessed May 9, 2018.

2

Chemists Who Work in Industry

2.1 Dorothy Jean Wingfield Phillips

FIGURE 2.1. Dorothy Jean Wingfield Phillips

Dr. Dorothy J. Phillips (Fig. 2.1) is a retired industrial chemist and a member of the Board of Directors of the ACS.

Dorothy Jean Wingfield was born in Nashville, Tennessee on July 27, 1945, the third of eight children, five girls and three boys. She was the second girl and is very close to her older sister. Dorothy grew up in a multi-generational home as both her grandmothers often lived with them.

Her father, Reverend Robert Cam Wingfield Sr., born in 1905, was a porter at the Greyhound Bus station and went to school in the evenings after he was called to the ministry. He was very active in his church as the superintendent of the Sunday school; he became a pastor after receiving an associate's degree in theology and pastoral studies from the American Baptist Theological Seminary. Her mother, Rebecca Cooper Wingfield, occasionally did domestic work. On these occasions, Dorothy's maternal grandmother would take care of the children. Dorothy's mother was also very active in civic and school activities, attending the local meetings and conferences of the segregated Parent Teachers Association (PTA) called the Negro Parent Teachers Association or Colored PTA. For that reason, she was frequently at the schools to talk with her children's teachers. She also worked on a social issue with the city to move people out of the dilapidated slum housing near the Capitol. The town built government subsidized housing to relocate people from homes which did not have indoor toilets and electricity. She was also active in her Baptist church as a Mother, or Deaconess, counseling young women, especially about her role as the minister's wife.

When Dorothy went to school in 1951, Nashville schools were segregated and African American children went to the schools in their neighborhoods. But Dorothy's elementary, junior high, and high schools were segregated even though the family lived in a predominately white neighborhood. This was because around 1956, and after Rosa Park's bus boycott in Montgomery, AL, her father, like other ministers, became more active in civil rights and one of his actions was to move to a predominately white neighborhood. The family was one of the first colored (the term then used) or African American families in that neighborhood. Despite living in that neighborhood, the children could not go to Murphy School as that was only for white students; they had to go to Head School which was for black children. Later the schools were integrated; her sister Elsie, who is ten years younger, did attend Murphy School.

The black schools in Nashville were very good from the standpoint of focusing on teaching and nurturing the students; Nashville has two historically black colleges and universities (HBCUs) and a medical school specifically for students of color: Fisk University, Tennessee State

University (TSU), and Meharry Medical School. Dorothy lived on the same side of town as her high school and the parents pushed for the schools to be academically strong because some of the parents might be professors at the colleges. The schools had college track classes with science and math and participated in the segregated contests for African American children. They also had an industrial track for those who were not going to college: beauticians, carpenters, and auto mechanics. For those who were being groomed for college, there were good math and science classes and dedicated teachers. The National Science Foundation (NSF) sponsored a summer science and math program for African American students in 1961 that was taught by the Chemistry and Physic teachers from Dorothy's high school. The program gave the students a chance to get hands-on experience. After she attended the NSF program one summer, Dorothy told her boyfriend, now her husband, about it and he went the second summer. They both became scientists. The next summer she was accepted at Bennett College for another NSF program. Her parents, concerned because they had never sent any of their children away for an extended period and Dorothy had been ill that spring, did not let her attend.

Dorothy had excellent teachers in high school. One she particularly remembers is Mr. Richard Ewing. He told his sophomore students that if they applied themselves, studied hard, and had good grades, they might get a scholarship to college. Unfortunately, Mr. Richard Ewing had a heart attack at the beginning of her junior year and died. Her chemistry teacher, Mrs. Laverne Holland, who had worked to bring the NSF program to the high school and taught chemistry in the program that summer, was also encouraging. Mr. Richard Harris, a physics teacher, made sure that she got scholarships to go to college. He helped her apply to the Tennessee Education Foundation Scholarship, which gave her funding to go to either TSU or Fisk University. She went to TSU, which was called Tennessee Agricultural and Industrial (A&I) State University at the time, because there she would not have any expenses as her scholarship and the work study program that she participated in gave her enough money. Her brother Robert, who was three years older, also won a scholarship to college and studied chemistry at Fisk University. When the family moved into their new, mostly white neighborhood, the other African American families living there were mostly teachers at either Fisk University, Meharry Medical School, or TSU. Living in such a neighborhood was great, because Dorothy's family was education oriented. These neighbors became role models for Dorothy and her siblings, encouraging them to get college degrees.

When Dorothy attended TSU she decided to major in chemistry and minor in math because of her participation in the NSF summer program. Before that summer, when she was fourteen years old, she had said that she wanted to go to college, major in history, become a lawyer, and then president of the United States. Now her plan was to go to Meharry Medical School to become a doctor after finishing college.

One chemistry professor that she knew at TSU was Dr. Rubye Torrey.[1] In later years, they connected at ACS national meetings and through Dorothy's daughter Vickie, who helped Dr. Torrey with the undergraduate Alpha Kappa Alpha Sorority, Inc. (AKA) chapter at Tennessee Technological University. Dorothy's analytical chemistry teacher was Mrs. Audrey M. Prather; Dr. Sadie C. Gasaway taught her advanced algebra. Both were members of AKA to which Dorothy pledged while taking their classes.

Dorothy left TSU after the first quarter of her junior year. She had made Who's Who in America for Students and pledged AKA. Vanderbilt University decided to integrate in 1964; African American students started as undergraduates there that fall. TSU chemistry major Max Collier decided to transfer to Vanderbilt University and asked Dorothy to join him. Rockefeller Foundation scholarships were available to the African American students. After discussion with her father, who was supportive, and her neighbors, Dr. and Mrs. Archer who taught at TSU and offered to write letters of recommendation, Dorothy applied. She was accepted and entered Vanderbilt in January 1966, still planning to be a doctor, because they had a medical school. She took the medical exam and discussed attending medical school with her advisor. One of the things he pointed out was, if she planned to marry, she should not go to medical school because he did not think having a family would fit in with attending medical school. He suggested that she go to Peabody College for Teachers (now a school of Vanderbilt) and get a master's degree in teaching to teach science. By then, she was engaged to her boyfriend. Remembering her participation in the NSF summer program, Dorothy decided to major in chemistry and minor in math. She took additional chemistry courses and got her bachelor's degree from Vanderbilt, then stayed for a year after graduating and worked in the psycho-pharmaceutical research department for two Vanderbilt professors, because her husband had one more year of college to complete at TSU. Working with biochemists spurred

Dorothy's interest in the field. But she also realized she did not want to be the person who killed rats for studies; she wanted to tell other people to kill rats. Hence, she decided to go to graduate school and applied to the University of Rochester and Tufts University.

Dorothy met her future husband in church. At that time, you didn't date just the boy, you dated the whole family. Her parents needed to know about his family, and finding somebody in church proved an excellent idea. Both she and James Phillips were in the church junior choir. He was a tenor; she, an alto. They were also in the same Sunday school class; they went to the Baptist Training Union on Sunday afternoon (which was called Baptist Young People's Union (BYPU), at the time). They started talking after church and found out they both liked math and were both going to be in the state math contest for Colored students. So, they started studying math together. They were teased at church because he is a year younger than she is.

Although they went to the same church, Dorothy and James attended different schools; he lived in South Nashville, where the church was located, and she lived in North Nashville. James went to TSU a year after Dorothy and majored in chemistry and math. Dorothy's transfer to Vanderbilt didn't hurt their relationship, because they mainly saw each other on Sundays, both at church and afterward when he would come to her house in the evening. His aunt made him wear a tie because he was going to Reverend Wingfield's home! It was a non-demanding boyfriend–girlfriend relationship, because they were really nerds and had a common interest in science and math contests. They became engaged in college and married two weeks after Dorothy graduated from Vanderbilt—on Labor Day weekend, September 2, 1967.

In 1968, Dorothy and James started graduate school at South Dakota State University (SDSU) in Brookings, SD. Her husband, a member of the army Reserve Officer's Training Corps (ROTC), had army duties on weekends and in the summer, but was exempt from the Vietnam draft. But it was cold in South Dakota, so they transferred to the University of Cincinnati, which also had an army ROTC program. Also, Dorothy's older brother Robert was a graduate student in the chemistry department there. Both Dorothy and James received teaching assistantships (TAs), which meant as a couple they had twice as much money as the average student. James did research in inorganic chemistry with Dr. Milton Orchin and Dorothy focused on biochemistry with Dr. Albert M. Bobst. James decided that he wanted to be in more of the business side of chemistry and needed an MBA, so finished his masters in chemistry while Dorothy continued to work toward her PhD.

The couple had their first child, Anthony, in March 1971 and their second, Crystal, in 1973.They managed parenting with tremendous support. Their neighbor Mrs. Martha Gray, who lived across the park from their apartment was a foster mother with the city of Cincinnati and became their babysitter. She took the children in the morning and dressed and took care of them until their parents came back from the lab in the evening. When James was working at Shepherd Chemical Company in Cincinnati, he took on the roles of both dad and mom by picking up the children and taking care of them after school. He wanted to make sure that Dorothy completed her graduate studies.

In graduate school, Dorothy's thesis adviser was trained as an organic chemist and had done his research on the measurement of biomolecules using electronic spin resonance, circular dichroism, and ultraviolet (UV) light. Dorothy concentrated on the R17 virus and understanding the conformation of the viral protein both as an entity with viral nucleic acid, which was a single stranded RNA, and isolated from the RNA. She conducted most of her studies using circular dichroism and UV light, with some using electron spin resonance. Her methods were slightly ahead of the times. Much of the conformation analysis then conducted involved using a computer and required punch cards. This kind of work is easier to do now because software programs are used to figure out protein conformational structures and changes, making data interpretation much less burdensome. Because her work might have had a link to cancer and viral diseases, Dorothy received financial support from Miles Laboratories. She has three publications about her research on RNA and its proteins in the R17 virus.

In graduate school, most professors expected their graduate students to go on to academia. Therefore, Dorothy had to teach a course to develop her teaching skills. While she was looking for citations for her publications she discovered the work of Dr. Patrick J. Oriel who was researching circular dichroism with proteins at Dow Chemical Company in Midland, MI. This sparked her interest in working at Dow Chemical Company. But this decision was not easy. In her marriage both partners were chemists and needed jobs.

Fortunately, Dorothy's brother Robert, who was already employed at Dow, gave the resumes of both Dorothy and James to the appropriate person and they were both hired. As an inorganic chemist, James worked in technical support for chlorinated solvents, while Dorothy worked in central research with Pat Oriel doing circular dichroism. Dr. Oriel thought it was a clever idea to hire her, because she was fresh out of graduate school and should have innovative ideas. Later at Dow, she got an opportunity to go into the chemical engineering group to work on a project to develop alternate energy sources to glucose and ethanol. While working on this project, she had more interactions with people outside of Dow. She went to the Natick Army Labs in Natick, MA, where the scientists conducted fermentations in their search for alternate energy sources. Working with this group, she became aware of raw materials that could be used for her own project at Dow. These interactions led her to realize that not everything has to happen in a lab, and she began to consider collaborations and interactions outside of her group. After the work with fermentations, her lab director decided that it might be worth studying how animals process varied materials. So, Dorothy got involved in animal research: How could digestion and metabolism processes be enhanced to speed up animal growth? This project, which was more in the realm of bacteriology or microbiology, was removed slightly from her field of biochemistry. The researchers wanted to know how to make the rumen digestion process in one of a cow's stomachs more efficient. They tested antibiotics and herbicides that were being used in screenings for new Dow products and identified microorganisms that produced nitrogen in soil and plants. Dorothy did gas chromatography on samples of rumen fluid in this search for animal growth promoters. The lab collaborated with Michigan State University, which has a degree-granting animal science program complete with farm animals. Students would bring rumen fluid to Dow twice a week and the research team used gas chromatography to identify the effects of antibiotics in the animal feed as growth promoters looking for enhancement of a certain marker. Dorothy got three patents in this area.

In 1983, James was offered a job at Corning Medical in Medfield, MA, managing the blood gas development laboratory. He decided to take the offer because he was working in the plant at Dow, not the inorganic lab. Dorothy, who was working in animal science developing animal growth promoters, wanted to return to research in biochemistry or perhaps, enter medical school. The family relocated to Massachusetts in the fall of 1983. As she was in the middle of getting an animal growth promoter approved by the FDA for Dow to commercialize, Dow allowed her to stay on as a consultant to complete the efficacy trials. From September 1983 to the following spring she frequently traveled back and forth between her home and Dow to complete the trials. When Dow decided to sell the rights to that drug and not bring it to market, Dorothy began her job search in the greater Boston area.

She connected with Frank McCarthy, a job recruiter and former priest, who helped her get interviews. His specialty was recruiting minorities for large corporations, and he facilitated her consultation at a new immunology project at Polaroid. She also attended meetings to network with people from different companies in the Boston area, including Millipore Corporation where she had an interview. Frank McCarthy had placed Dr. Joseph C. Hogan Jr. from the Polaroid Corporation in a position as vice president at Waters Corporation. Frank told Dorothy that he would send her resume to Joe after he joined Waters, because he would most likely want to bring in his own staff. He did, and she was brought in to develop chromatography packing materials for proteins, peptides, and nucleic acids. Her department was called Chemical Research and Development (CRD) at Waters, but it was a small department. Waters was more focused on developing instruments than consumable products. Chromatography originated as the science of separating the components of a solution and measuring them by their assorted colors. Today detection of components is not limited to the use of colors and there are many applications of chromatography. For example, in clinical trials to determine how the body has metabolized a drug, new components or metabolites will be noted in the chromatogram, and this analysis helps to determine the most efficacious level of the drug.

At Waters, Dorothy worked in a new group in CRD, bioseparations. The feeling was that her background in protein and nucleic acids could help them move into a new area. Dorothy's objective was to help develop the chemistry in the tube or column so it would perform well with large not small molecules and not with polymers but biomolecules that were related to the body—proteins, peptides, and nucleic acids (such as DNA). Millipore Corporation also wanted to get into the biotechnology industry. Waters, a small company started by James Waters in 1958, had been bought by Millipore in 1981, as by then it had grown and had a focus on chromatography. In 1984, when Dorothy started there, it was a subsidiary of Millipore, which wanted Waters involved in their transition into the new and booming biotechnology industry. Therefore, Dorothy became a part of the biotech consulting group under Jack Mulvaney (Chairman, President and CEO, Millipore, died in 1986) in Bedford MA. She was one of the consultants developing a plan for Millipore, which gave her greater visibility and familiarity with the staff in the Bedford office. Her group was successful in developing a product using the instruments designed for purifying proteins, the family of AccellPlus™ packing materials, which could purify two distinct types of proteins. Dorothy's team published the research and applications after it was released to the market.

In 1989, Dorothy was invited by the People to People International Program to go for a three-week trip to China as part of a Western delegation. She had never been to Asia and she was excited, even though she had never been away from her family for that long. But, the trip was delayed a year because of the Tiananmen Square incident.[2] The group, when it finally traveled, included speakers from Princeton University, University of Chicago, Washington State University, and pharmaceutical companies. They spoke mainly in universities, pharmaceutical-type companies, and pharma schools. Dorothy submitted her paper a month ahead of time for it to be translated, but for some protein names, like lysozyme, there were no translation. She flew first to Millipore's Hong Kong office—the trip was funded by Millipore because of its interest in this emerging country. After meeting with Millipore representatives in Hong Kong, she joined the rest of the group and toured China, giving talks and sightseeing. The host for the trip was the Chinese Academy of Science and Technology (CAST). That was her first trip to China, but she traveled there more often after joining the marketing group at Waters.

Working in R&D, Dorothy was involved developing AccellPlus™ sorbents and a suite of other products for the developing biotechnology market. Her research later changed from analysis of large molecules to analysis of small molecules, including drugs like Viagra and Lipitor. In doing so, she got involved in developing columns and chemistries that performed better than the current technologies did at the time. That work and her publications on the work led to more recognition, and she was invited to speak more often to the scientific community outside of Waters. Dorothy was interested in the applications of the products she had helped develop and was traveling to give seminars across the country and in Europe. Because of that, she was given the opportunity by Waters to go into marketing and was encouraged to do so by her son. She became a brand manager for a product that she had helped to develop which was, Oasis[R] sample extraction sorben. She also trained Waters staff and gave seminars at pharmaceutical companies worldwide which was part of her marketing work. In her role as a strategic marketing person, Dorothy traveled even more frequently, sometimes to understand unmet market needs, especially in Japan and China. China had water-testing and environmental issues and Dorothy needed to know if they had the proper tests to monitor such issues. This was to help them find the proper instruments to do the job. In 2004, the People to People International Program invited her to return to China. China had water testing and environmental issues and she needed to know if they had the proper tests to monitor such issues. This visit focused on the pharmaceutical industry, Western medicine and traditional Chinese medicine (TCM) companies, because of the many components of the mixtures in traditional Chinese medicine, she learned there was a need for analytical chemistry as a result, Waters got involved in TCM analysis, which helped the organization grow in China. Leading up to the 2008 Olympics, Dorothy visited the Centers for Disease Control (CDC) of China and the provinces of China to understand what they needed to analyze food and water samples. Waters did not get all the business that came out of this trip, but they got part of it.

When Dorothy had transferred in 1997 from research into marketing, both her children, Anthony and Crystal, and her stepdaughter Vickie were out of college. Her stepdaughter is her husband's daughter who lived with them from junior high through high school and is now married and an integral part of the family.

Dorothy's responsibilities at Waters increased when she went into marketing. In November 2012, she made a farewell retirement trip to China and Japan. She was given wonderful retirement parties in the Waters offices in Shanghai, China, and in Shinagawa, Japan. She received many gifts and felt appreciated because of what she had done to help the company grow the business in Asia. She had also traveled to India, Korea, Taiwan, and Thailand to give seminars or meet with customers.

When she retired, the multiple responsibilities of her job were divided among different people. Although she had planned to retire in September 2012, the company did not identify the person to handle a major portion of her strategic marketing activity until January 2013. Since the candidate needed training, Dorothy stayed on until March 2013. Even after she retired, she was available as needed. In summary Dorothy was one of the few African American women chemists who could begin her career in industry as a senior research chemist and end her career after thirty-nine years as a company director. Dorothy went from just working in the lab at Dow to becoming a leader in the lab at Dow to working for Millipore which was bought by Waters. There she was asked to work as a leader in several departments, so she went from the lab to the board room. This was very unusual for an African American woman chemist.

When asked what, she would say to young women who would like a career in industry as a chemist Dorothy said: "No longer will they stay at a company for thirty or more years. However, in that brief period [they will be there] they must plan how they can contribute to the company's bottom line and their career. They should make sure that they do something that helps the company earn money or increase its revenue or bottom line. Therefore, if they stay at the company for only one or two years, they can say that they contributed. To do that you need to be on a good team and understand what your manager wants. They could also set a course to stay in industry for a few years to study how to become an entrepreneur. They should set goals for career, finance, social life, and personal life. They should identify organizations that are beyond their local area. Also set spiritual goals, such as teaching Sunday school or serving in an organization and practicing personal devotion. You will also need a mentor."

When asked about mentors she said that when she came to Waters, Dr. Joseph Hogan was head of her department and also the reason she got an interview. He was not her direct manager but had said to her that there were people with a 1940s mentality as far as people of color were concerned. He wanted her to understand that and know that she could talk to him if she experienced it. She did not have this experience because she felt it is often your attitude toward the person. Lashing out at the person helps no one but if you are both civil to each other it is okay. But she was happy to know that Dr. Hogan was there to make sure that things were going well for her. He was there for her if she had a problem with the attitude of other people toward her because of her race.

Also, John Nelson, who later became the senior vice president of R&D and chief operating officer for Waters, took an interest in her and fought some battles about what roles she played in the company during the early years. He encouraged her to take on tasks. For example, on attending an ACS Division of Analytical Chemistry meeting they realized that other companies sponsored awards. He said to Dorothy that Waters did not sponsor an award, but he thought they should. She did not take this statement as an invitation to take on a task until he asked about it again. Waters now sponsors two national ACS awards and until 2014, sponsored an analytical division award. Dorothy and Dr. Martin Gilar, a senior level scientist, went to senior management team to establish an Uwe D. Neue Award—Dr. Uwe D. Neue was from Germany and excelled in the analysis of small chemical structure by chromatography. But he had less training in the analysis of large molecules like protein and nucleic acids, which Dorothy taught him. In turn, he taught her about small molecule analysis and expected her to publish every year and present at conferences. He not only expected it, he made it happen. More than a manager, he was also a colleague working directly with her, and ensuring publications were completed. He promoted her to manager and discussed the direction of the group with her. He was well known internationally, from both publishing books and speaking worldwide. They both traveled to China in March 2010 to give lectures and help develop collaborations. His death in December 2010 was a great loss for her.

Ian King, director of marketing and later vice president of consumables business taught her marketing. When Dorothy became brand manager for the Oasis^R family of sample preparation sorbents, he traveled with her, demonstrating firsthand how to train staff on a new product and encouraged her to sign up for marketing classes. They traveled the world together; he helped her learn the business side of chemistry. In her first year, sales far exceeded the forecast for the first two Oasis^R products.

Her last mentor before she retired was Michael Yelle who respected her expertise and would help with resources needed to get a task done without question.

As a mentor, herself, she supported the people in her groups, especially the young ladies. They could come to her, she would close her door, and listen to their concerns or inquiries and send them out on a positive note. So, mentoring works two ways.

When asked about how to succeed in a large company she said, "You need to begin where you are. Study the company and see what you can do to achieve something there. There is really nobody that you cannot approach; but do it in the right way with proper preparation. Find out who makes the wheels turn in that company and what your role should be. When you see opportunities, first speak to your manager about it. Write up everything you do in the form of a report and think of who to send it to. This may be more than your direct boss, anyone from colleagues to directors. Learn to communicate well. This can be done by joining Toastmasters International. Remember, you need an edge that can move you forward. You are not trying to best your colleagues, but putting your best foot forward. Learn to be a team player; you must work well with the team. Also, think outside the box. Join organizations both national and local, like the American Chemical Society, ACS."

Dorothy's involvement with the ACS started when she was at Dow Chemical. A big company in a small town, Dow expected their R&D people to participate in such organizations. At first, she was involved in only the ACS Midland Section, at the local level where they organized symposia, meetings, and other events.

In Midland, there were only a small number of African American scientists. So, the Midland chapter of NOBCChE gave them an opportunity. Dorothy became vice president of NOBCChE in Midland. In addition, because there were very few minorities in Midland those who were there wanted to have an organization that would give their children an advantage and foster their interaction with others. The Midland Black Coalition offered several programs and brought a predominantly African American Greek organizations to Midland, like the AKA of which Dorothy is a member.

In the Boston area, Dorothy joined the ACS Northeastern Section (NESACS) and participated in Project SEED, chairing the committee and obtaining funding from Millipore, Waters, and Polaroid, among others Project SEED is an ACS program that places economically disadvantaged high school chemistry students in summer jobs for which they are paid by the ACS. She identified teachers willing to take students into their labs and work with the program in the Boston area. When she was asked to run for chair of the ACS local section she did so. However, she did not think she would win because the person running against her was more well known in the Boston area than she was. Chosen chair-elect in 1992, she took the chair in 1993 and attended the national ACS meeting in Denver to present the ACS National James Flack Norris Award for Physical Organic Chemistry sponsored by NESACS. Dr. Esther A. H. Hopkins[3] told her what would happen at a national meeting. Esther took her to the Board-Staff reception that was hosted by John Crum, Executive Director of ACS (1982–2003) and introduced her to many people. She advised Dorothy to visit different committee meetings to find out what was happening. Dorothy went to the Committee on Minority Affairs (CMA) meeting, where she saw another side of the ACS. When she finished her job as chair she was on the NESACS nominations committee and was asked to run for the ACS national council. She became a councilor in 1995 and attended her first council meeting in Chicago, where she also visited committees. Although considering participation in the CMA, she was appointed to the Membership Affairs Committee. Locally, she was still on the Project SEED committee and chaired both the fundraising and awards committees. Nationally, she was elected to the Committee on Committees (ConC) and served two terms. While on ConC she was liaison to the Younger Chemists Committee and to the Nomenclature Terminology and Symbols committee. She was appointed to the Committee on Divisional Activities and then was elected to the Council Policy Committee where she served two terms.

Working with the ACS Division of Analytical Chemistry to set up the awards for Waters to sponsor, she got to know the officers and several members of the division and got more involved by working on tasks and project teams. She ran for chair of the Analytical Division in 2008 and won; it was a four-year term. She ran for Region 1 Director and for director-at-large and in 2009 was asked to run for ACS president but she did not win. She felt that she could bring a lot to the group with her global experience and could help the organization expand its reach globally. Since 2013 she has won two terms as director at-large on the ACS Board of Directors (2014–2019).

Dorothy served on the 2012 Presidential Task Force "Vision 2025: Helping ACS Members Thrive in the Global Chemistry Enterprise" while Dr. Marinda Wu was president. She got to know Marinda because they both had worked at Dow. Dorothy received input from several committees to develop task-force recommendations for the Board. Dorothy gave a talk at an ACS national meeting on active learning and wrote one chapter in an ACS symposium book. She said young people should do internships while in college, which would be a part of their education. These internships should not only be local but global as well.

In 2006, Dorothy received one of the Unsung Heroine Awards from Vanderbilt University, because she was the first African American woman to graduate from the university. The women recognized that evening also included students from Peabody College of Education and Human Development, a school of Vanderbilt. Dorothy gave a talk titled "Crossing the Road" at that event. Afterwards the Director of the Bishop Joseph Johnson Black Cultural Center approached her about establishing a leadership scholarship in her name. This award is named the Dr. Dorothy Wingfield Phillips Award for Leadership and is presented to a graduating senior who has demonstrated strong leadership skills.

Dorothy was asked in 2004 to speak at the Nashville ACS section, which was chaired by her brother Robert (Dr. Robert C. Wingfield Jr., Fisk University), and she received the Nashville Section ACS Salute to Excellence Award. Many students were at that meeting from Fisk

University, Vanderbilt University, Middle Tennessee State University, and Tennessee State University. In 2011, she was invited by the New England Institute of Chemists to be a mentor at their Brandeis University program for students.

Dorothy is a part of the TTT Mentor program in Cambridge, MA, a program started by one of her sorority sisters to give back to the community. Kathleen Jones who started the program has a background in science. She has an ongoing program on Saturday morning where children, in grades four through seven and even younger, work with mentors and develop science projects that are showcased at a fair at MIT each year; the posters describing the projects are judged. She receives funding from the City of Cambridge and is seeking additional support. The mentors are mostly graduate students from MIT, Harvard University, and other colleges in the area. These young children take the ideas they get from their mentors and work with them on the projects. Now that she has retired Dorothy hopes to have more time to work with her husband and the students on Saturdays.

Although Dorothy became a member of AKA as an undergraduate student at TSU, she really began to work with them when she was in Midland, MI. She felt it was good for her children, especially her step-daughter, Vickie, who needed to build her self-confidence. In Midland, there were few minority children. Bringing them together as a group gave them an opportunity to show that they could be leaders.

Her work in Boston with AKA was different. She wanted to be involved in programs that would help the minority communities in Roxbury, MA, which is a predominantly African American area. From 1994–1998, the sorority ran a program called Partners in Math and Science that caught her interest. Dorothy asked science teachers in the Boston area to join a committee to work with students on Saturday mornings. Students entered a local contest and the winner went to the regional AKA meeting and the regional winner went to the national meeting. In that program, a young lady from Boston went all the way to the regional level in the science contest. In addition, one of the student sorority members Dorothy worked with was a chemistry major at Wellesley College—Malika Jeffries El is now a PhD chemist active in the ACS. The sorority gave Dorothy an opportunity to give back to the African American community. She has been on the scholarship selection committee several times. It also holds debutante balls to raise money for scholarships.

Dorothy is a member of NOBCChE, but the Boston chapter has been quiet since the demise of Polaroid. The group is trying to reorganize. In the past they ran a science fair at the Boston Museum of Science and had a Saturday science workshop for sixth graders.

Dorothy now has three grown children. The oldest, Vickie, went to TSU for a degree in business administration and Tennessee Tech for a master's degree. She works in the computer field, implementing PeopleSoft software programs in universities, for IO, an organization that implements PeopleSoft software in different settings. Vickie married her college sweetheart, Albert Thomas, and they live in the Atlanta Georgia area where her husband is on the faculty and coaches at North Atlanta High School.

Dorothy's son Anthony, the middle child, attended a private high school, Rivers Country Day in Weston, MA, so that he would develop fully academically. He went to the University of Pennsylvania, where he played varsity lacrosse, and made both the Lacrosse Hall of Fame as well as All Ivy and became a Senior All Star. He graduated with a major in economics and a minor from the Wharton School in Marketing. With colleagues from Penn, he set up a business called Persaud Brothers marketing and advertising. In Atlanta, Georgia he developed a marketing campaign for Bellsouth Corporation. He was married to Brandi Williams whom he met in Atlanta. He has an MBA from Emory University and worked for Coca-Cola for six years in global marketing. He received his Master's of Divinity in May 2017 from the Candler School of Theology at Emory.

Dorothy's youngest child Crystal went to Ithaca College. She was a swimmer in high school. At Ithaca College, she studied business accounting. She was married to Stephen Mayo who attended Framingham State University and is a hardware engineer in the computer industry. Crystal first worked with Stride & Associates in Boston but changed jobs to be closer to home after her children were born. Her daughter is now a financial analyst at National Grid in Waltham, MA.

Dorothy thinks she is blessed by having had her children while she was young. She and her husband pushed academics with them. Since Dorothy came from a strict Baptist home, there were strong rules to adhere to. Dorothy and James worked with the children in church, teaching their Sunday school classes. The grandchildren have been baptized, which is important to Dorothy. Since her father was a minister, Dorothy had thought he had the strongest role in her family. It wasn't until her father died in 1981 that Dorothy realized her mother was also strong. Her mother nurtured all her children and grandchildren, but especially her grandsons. Everyone came to her house on holidays. She felt it was important to work with the boys because life can be very tough for African American men.

Although they are intelligent and educated, they need more. Dorothy hopes to be the same kind of grandmother, one with interests, who imparts her wisdom to the next generation. Her mother died in 2010 at the age of 94.

Now that Dorothy has retired she hopes to write her autobiography. In 2013, she had an experience with her grandson Andrew Phillips who lives in Atlanta. He was nine years old and did his Black History Month project about his grandmother, Dorothy. His teacher spurred an interest in the students about the Civil Rights Era and sent out a note to the parents asking if any of their children's grandparents could come to the school and talk about living through that time. Dorothy's son sent the letter to her, asking her to speak. She decided to go with her husband and they both talked to the third, fourth, and fifth grades. When she told them that she had heard Dr. Martin Luther King Jr. speak one boy asked her age because he thought that happened ages ago. Since they asked so many questions she feels she must write the story and offer it to other schools as a program for their curriculums.

When asked about growing up in a Jim Crow state, she reflected that as a child your world is only as big as what you are exposed to every day. Until she was about eleven her world was African American schools and churches; her neighbors were a group of people who owned their modest homes. The few white people she knew employed her mother. Her world was contained. It was not until the family moved to Murphy Avenue that she found the world is not always nice to you and some people don't want you around. The neighbors called the police when she and her siblings went out to play. Her older brother was active in the Civil Rights Era, participating in sit ins and getting beaten up once. She remembers segregated drinking fountains. If they wanted to see popular movies they had to go around the back of the downtown Nashville theaters and go upstairs to the seating for Blacks.

When things started to change in the 1960s, they began to realize that what they thought was normal was not. Change was the tough part, moving into a world with people who did not always want you. When she went to Vanderbilt, it was the first time she went to school with whites and had white professors. It was tough because she felt lonely and sort of rejected. People were not always supportive and looking out for her best interests. But she learned to rely on her own inner strength and faith.

Growing up in the south there were limitations that she did not realize until she was older. Their school books were not the best. But that did not stop teachers from giving their all to the students because they knew what they would be up against in some ways. In some aspects students fared better in the segregated schools than her children did in the majority schools. For example, her youngest daughter went to a school where students had been bused from Boston. Some teachers assumed all the students of color came from Boston and did not want to support them. One teacher thought Dorothy was the parent of one of the bused students and was surprised to learn they lived in the neighborhood. Dorothy is grateful for her parents who pushed their children to be what they could be. Not one of her siblings had a sense of inferiority about being African American.

When speaking about diversity and employment she said Waters did not embrace diversity in her early years in the company; the company said that its diversity reflects the community, which is not very diverse. When she retired, as a director she was the highest-level African American woman in the company. But Waters is gradually changing. There are now more people of color than there used to be. Dorothy was still the person that fellow minorities would go to and talk with. She had spent time just before she left with one young lady who was originally from Africa. She wanted to know how to get into leadership at Waters. She had been there two years and wanted to be a manager.

Overall, she feels that she could move forward as her work and her achievements merited. Women are not going to be where men are in the company, but she challenged them on that. They still need more women in leadership roles in that company, but it's coming. They have one lady on the executive staff who is a senior vice president, but there should be more. There is still a glass ceiling.

Dorothy is now a lay leader for Fisk Memorial United Methodist Church, having left the Baptist Church with her father's blessing—during graduate school the church in her neighborhood was an AME (African Methodist Episcopal) Zion church. As lay leader she does what the pastor requests for programs, ministries, and helping members and covers when the pastor is absent or on vacation.

Dorothy is grateful for the women in her life who helped her. For example, Dr. Esther Hopkins, a chemist friend who rode to sorority meetings with her on Saturdays. She encouraged and inspired her by giving her life lessons, like not buying a used car because you get somebody else's problems. Another sorority member, Bertha Wormley, taught her how to drive in Boston and introduced her to the inner-city community. Dorothy reciprocated by driving Bertha from Natick to Boston for AKA meetings. Her babysitter while in graduate school at the University of Cincinnati was

Mrs. Martha Gray; she and her husband took care of her family at the critical time when they were finishing graduate school. Dorothy dedicated her thesis to Mrs. Gray and her husband because they enabled her to finish.

Dorothy calls her husband her rock. They went to graduate school together and he encouraged her to get her PhD. With each baby, he encouraged her by giving assurance, ensuring there would be housekeepers and babysitters to give her the space to do what she needed. They have been married since 1967 and there are no bad stories. He does not lie, not even to trick the children. He expects a lot of their children and of himself. They worked together at both Dow Chemical and Waters. To gradually transition to a new reality, he did not retire when she did.

She sums up her life with, "I'm not done yet."[4]

2.2 Cherlynlavaughn Bradley

FIGURE 2.2. Cherlynlavaughn Bradley

Cherlynlavaughn (Cherlyn) Bradley (Fig. 2.2), who was born in Chicago, IL, became an industrial chemist and vowed early in her career to help young people, especially girls, follow in her footsteps.

Cherlyn was born on October 27, 1951, to Leroy and Geneva Bradley. Her father, who had a tenth-grade education, worked as a carpenter and freight handler for Burlington Railroad (Chicago, Burlington, and Quincy Railroad) and later attended Dunbar Trade school in Chicago. Her mother graduated from high school and also had some college education; she was a musician, piano teacher, and interior decorator. Cherlyn was their only child. The family lived in Chicago, first on the south side, then the west side, and finally the suburbs.

Cherlyn was educated at St Matthew Catholic School (St Matthew Parish School) from grades one through five. Her parents enrolled her in the private school because, since they lived on the south side of Chicago, African Americans could get a better education in a private school. Her family moved to Broadview, IL, to get out of the city. Broadview is a small town, which at the time was not integrated. Although her mother still wanted her to attend Catholic school, there were no openings at the time. So Cherlyn attended Roosevelt Public School for sixth to eighth grades. When she graduated from Roosevelt Public School, she went to Proviso East High School, a regional public high school in Maywood, IL, which is next to Broadview. Her high school was integrated and had good teachers who cared about her.

Cherlyn first became exposed to science at home. She saw her first microscope during the fourth grade. Then her parents bought her a Gilbert Chemistry set in the fifth grade. Since she was an only child she frequently got both girl's and boy's toys. And she quoted Shakespeare, "I am all the daughters of my father's house, and all the [sons] too." She was not interested in dolls as much as trains, baseball bats, and basket-ball hoops. In fact, she played basketball with one of the neighbor boys because she had a hoop. She set up her Gilbert Chemistry set in the attic room of her house and had a Bunsen burner in the utility room. She did not start any fires or blow anything up but an interest in chemistry was triggered.

Her science courses started in the fifth grade. In the sixth grade, the class visited the high school and saw the labs. She loved the chemistry lab, since she already had her chemistry set. When she attended that high school she wanted to concentrate on chemistry. After looking at the course list, she decided that in order to take an accelerated chemistry course, she would have to take general chemistry in the summer after her freshman year, and did so. The teacher of that summer course, Mr. Isley, became a great influence on her career. In high school, she took all the chemistry courses that were offered, including an Advanced Placement class. But she also took non-science courses that would help her in her future career, like history and debating. Her choice of college was Illinois Wesleyan University in Bloomington, near the center of the state, because a representative of the college came to the high school to recruit, chatting with some of the students during career day. At the end of her junior high school year, she took a college level course in creative writing there and met some of the college students, other high school students who were also taking the course, and the faculty. Feeling already part of the place, she was set to enter college there and enrolled upon graduation from high school. The college was integrated then, and even more so now. But it was a long way from home, and Cherlyn lived in the dorm with a roommate. While Cherlyn is Roman Catholic, her first roommate was Baptist. They got along, but the next year her roommate wanted to room with a fellow Baptist student. So Cherlyn met a new roommate, also Catholic, who remained a friend.

Since she had already declared a major in high school she started chemistry in her freshman year. She also took the required freshman-year courses in history, anthropology, English, and so on and elective courses in music and criminology. For her major, she took all the chemistry courses available: research, physical chemistry, organic chemistry, and synthetic organic chemistry, and did some research in organometallic chemistry, mostly in the library. Most of the chemistry that she did in college was inorganic because that was the professors' interest. At the time, she was possibly the only African American woman chemistry student. Her mentor was Dr. Wendell Hess. Cherlyn graduated in June 1973 with a bachelor of arts in chemistry, magna cum laude. Then she decided to go on to graduate school and major in inorganic chemistry with a goal of a future position in industry. Her choice of school was aided by her college counselor, Anne Mayrhofer. When asked where she would like to go to graduate school, she said she would like to remain in state. So, her counselor gave her a list of universities in the state that did research. Northwestern University was on the top of the list, so Cherlyn decided to apply there. It was the only place to which she applied. She received a teaching assistantship and entered in the fall of 1973.

While she was at Northwestern, Fred Basolo, a famous inorganic chemist, was the chair of the department. At the time, Cherlyn was the only African American student in chemistry until the year she graduated.

At Northwestern, in order to get a PhD, you first had to get a master's degree. In order to get a master's degree, you had to do a virtual research project. This involved thinking about the project and putting it on paper, including proposing a project, creating a hypothesis, and proposing the steps needed to solve the question and get to the conclusion. The project was submitted it to the professors and if it worked, you got your master's.

To go on to get a PhD, Cherlyn had to take qualifying exams in physical, organic, and inorganic chemistry. Anyone who passed those exams did not have to take actual courses. Since Cherlyn was a TA, she had to help teach undergraduate students and used that opportunity to review her organic chemistry for the qualifying exam. But Cherlyn felt she was weak on physical chemistry, so she took courses in that and in organic chemistry.

After successfully completing the exams, Cherlyn started to do research in order to write her thesis. First, she had to pick a thesis advisor. Cherlyn looked at the work of the professors in the inorganic chemistry department and chose Dr. Allred Marks and to work in the area of synthetic inorganic chemistry. The long-term goal was to make a semiconductor.[5] The short-term goal was to synthesize as many of the structures as possible and test them to see if they would work. As Cherlyn said, it was the professor's project. She conducted this research from 1974 to 1977, working on the preparation of covalently attached permethylpolysilane groups to platinum electrodes by electroreduction of chloropermethylpolysilanes, on characterizing chemically bound products using x-ray photoelectron spectroscopy and NMR, and on the synthesis, separation, and characterization of two new cyclic silanyl oxides as potential inorganic polymers and semiconductors. The goal was to make a polymer consisting of silicone with methyl links similar to an organic polymer. She successfully defended her thesis in October 1977 but did not get her degree until the graduation ceremony in 1978.

While at Northwestern, she was inducted into Alpha Gamma chapter of Phi Lambda Upsilon, an honor society for graduate chemistry students.

In 1977, the Civil Rights Act had only been in place for about twelve years. Science recruiters from Bell Labs and Standard Oil came to Northwestern in the spring looking for chemists. Cherlyn liked the story that Standard Oil of Indiana was telling and decided to seek a job with them. She interviewed and was accepted for a position! But she was still writing her thesis and had not yet defended it, so she could not begin for them right away. In fact, she received letters asking her when she was coming, which she took to mean they really wanted her. She didn't start her job until October of 1977, even though she had been hired six months before. Standard Oil of Indiana has changed names several times becoming Amoco Corporation and later BP Amoco.

Her job at Standard Oil was in Naperville, IL. She was hired to work in Analytical Research and Services doing gas chromatography, element-selected detection for all the fragrant sulfur compounds. The lab had a refrigerator loaded with the bottles of mercaptans and sulfides and disulfides. Since the refrigerator smelled like rotten eggs from the mercaptans, workers put the compounds in a ventilated hood and got rid of the refrigerator. They were interested in analyzing trace contaminants in petroleum and in petroleum products, from crude oil on up. Of course, it's more than trace in crude oil. But by the time you get to a finished product like gasoline, government regulations [stipulate] that you need to only have a certain part per million levels of sulfur-containing compounds in there. For example, in natural gas which people have in their homes, sulfur products are put into the gas so that people can smell a gas leak, but in gasoline you do not want to have a smell. Her job expanded to testing for nitrogen and oxygenated products as well.

Because of the work her lab was doing, Cherlyn was asked by her boss to join the American Society for Testing and Materials (ASTM),[6] a group of representatives from different industries who do cooperative laboratory work to standardize tests for a host of things. And among those items was carbonyl sulfide in polypropylene, which was the research and analysis that she perfected with her group at Amoco at that time, and she wrote the documentation. She was told by her boss that working with ASTM would help her career at Amoco. So she joined ASTM and chaired the gas chromatography[7] study group, representing Amoco. She was able to analyze for carbonyl disulfide (COS) and propylene. Gasoline providers want the minimum of those two products in the gasoline so they coordinate on the standard.

The gas chromatography study group involved collaborative work between Amoco, Royal Dutch/Shell, Chevron Corporation, Mobil Corporation, Marathon Oil Company, and overseas companies which attended the ASTM meetings. So, her COS and propylene study group had an attendance of fifty to eighty people. They would bring samples to be tested and the group would agree by consensus on a standard for all petroleum companies even some small ones not involved in the group.

Back at Amoco, Cherlyn advanced to a senior chemist position managing a small team in analytical chemistry. The team consisted of technicians with a two-year degree who might be working on a bachelor's degree to become chemists.

Because she worked in the field of element-selective detection, Cherlyn was able to interact with the refining group at Amoco as her client. This required that she go into the field, and so she spent about three months at Whiting Refinery in Indiana, because they wanted to set up their laboratory using the procedures from Naperville and the ASTM standards. Although Cherlyn was the consultant, she was actively involved in setting up the analytical group. At Amoco that's the way it was. Chemists would get into the laboratory, set up instrumentation, and work shoulder-to-shoulder with the technicians. It was more than just doing deskwork and delegating, although that's done too. Cherlyn also traveled to refineries in Texas where the lab did oxygenate and sulfur work and to Chocolate Bayou Refinery where she set up instrumentations and helped develop methods for sulfur-containing.

As BP merged with Amoco, Cherlyn still did laboratory work, but she was also quality assurance coordinator—later manager—for the Analytical Division. Later the analytical group was divided up, so there was no set analytical services group in BP. At that time, she had PhD chemists reporting to her for a while.

During her career at Amoco she did have mentors—Jeff Meyer and Ken Albert. Jeff Meyer taught her how to change from creative writing to scientific writing, which is important in industry and for publications. He passed on what another had given him. Ken Albert helped her develop her customer liaison capabilities and helped her become familiar with his technical analytical area, which she worked in upon his retirement.

There were few women (and she was the only woman of color) in her department at the time. So, she was mentored by men. In general, this was the rule at the time in Amoco. She knew of women chemists in the Analytical Department and maybe some women chemical engineers in the Refining Department. Cherlyn might possibly have been the first African American Woman PhD chemist to work for Amoco. She made it a point to mentor young women chemists when they came to work for Amoco.

Cherlyn was involved in improving the design of an instrument called the Microwave Plasma GC Detector, which detects different elements in a sample by printing out differently sized peaks which are the amounts of the chemicals and may be only used in the petroleum industry. The company bought a prototype machine, as big as a refrigerator, from London. When she discovered that Hewlett Packard was developing a smaller machine, Cherlyn told them what its users would require. Since she had done this collaborative work with Hewlett Packard, she was able to be at their booth at the Pittcon Conference to talk to people about the instrument.[8]

Working in industry, Cheryl found it was hard to write publications for journals and to give talks at conferences or to the public. Her clients at ASTM told her the work was private and should not be published. Therefore, the publications and the talks that she did produce (and are listed here) came out long after the work was done.[9] Scientists, even industrial scientists, are judged on their publications and talk which helps them for promotions or to get another job.

When BP and Amoco merged in 1998 they began offering retirement packages. After reviewing the packages, Cherlyn decided in 1999 that it was a good time to take early retirement.

She did not regret the decision and even in retirement has kept active. She joined the Volunteers Retiree Group, which started at Amoco in 1992, is supported by the BP Foundation, and is active in many different causes.

The one program Cherlyn chose to work was with Court Friends. In her county, many disabled people are wards of the court with appointed guardians of some sort. Volunteers check up on the wards to see that they are okay.[10] She also worked with Friends of the Little, or Silent Santa, at Christmas in which they donated gifts to children. Not all the retirees in the group are chemists. Once the chemists in the group did demonstrations for children in the museum.

Cherlyn was also active in her local Catholic Church and a member of the Parish Council. This group consists of four commissions that rule the church: liturgy, human services, education, and help each other. The council also gives input to the Archdiocese of Chicago on the hiring of new priests. Cherlyn served as chair of the membership and nominating committee, led a Bible study group, served as church library chair, and was a Eucharistic minister. She also served as member and Secretary of the NorthStar Credit Union Board.

Cherlyn chose never to marry. In addition to her volunteer activities she also enjoyed many other things. Although her mother taught her to play the piano when she was young, she gave it up to study science, and got into sports, including tennis, baseball, basketball and golf. She also enjoyed spectator sports, especially baseball and basketball.

Cherlyn was an active member of the Chicago section of the ACS and also the national ACS. She became a student affiliate member of the ACS as an undergraduate at Illinois Wesleyan University, joined the ACS in 1975, and became active in the Chicago section in 1980. She joined a number of committees including Project SEED, which brings economically disadvantaged students to work for a summer in an academic, government, or industrial lab. The local section had two or three students a summer, it was not hard to find students, it was harder to find mentors, but they worked on it. Cherlyn and her friend Fran Kravits also mentored girls for a local Girl Scout chemistry award. She did not think the Girl Scouts had a national award.

Her other roles in the ACS section included being treasurer, vice chair, and chair of the section. She chaired the analytical topical group, and edited the newsletter, the *Chemical Bulletin*, through ten issues a year. This was a big job, since she had to assemble all the data and coordinate with the proofreader and the webmaster for display. The newsletter was sent electronically to all members.

One major event for the Chicago and other Illinois sections is the science tent at the Illinois State Fair every August. Thirty to forty volunteers do demonstrations and lead activities, including computer-based activities, for kids of all ages and for teachers. Teachers get professional credit for visiting the tent and a small gift when they leave. The fair runs for ten days and volunteers are there the whole time. A committee plans the event and raises funds for it.

Cherlyn was elected councilor from the Chicago local section to the ACS national meeting. A councilor is like being a congress person in the House of Representatives. (The ACS Board of Directors are like the US Senators). She started as an alternate councilor in 1991–1992 and served as full councilor from 1993–2014. As councilor, Cherlyn was expected to serve on a committee of the ACS. She was active on several—Project SEED, Meetings and Expositions, and Chemical Safety—and chaired the Partnerships Subcommittee. She was also elected to the ConC, where she was the liaison to several other committees, including International Activities, CMA, Women Chemists Committee (WCC), Committee on Nomenclature and Symbols, and the Joint Subcommittee on Diversity and Inclusion. The last committee to which she was elected was the Committee on Nominations and Elections.[11]

Dr. Bradley's final advice to students is, "Try to be as well rounded as you can. That is, take the science, take the history, take the English, and get involved . . . get positive in order to stay positive because there are opportunities, even though there may be difficulties. It's your generation and it's your chance. You can succeed." She said she came from a religious background, and so she added, "with God's help."[12]

Cherlynlavaughn Bradley, 62, retired BP senior research scientist, died of cancer, on August 30, 2014.[13]

2.3 Sharon Janel Barnes

FIGURE 2.3. Sharon Janel Barnes

Biologist and chemist Sharon Barnes (Fig. 2.3) created both the process and apparatus for contactless measurements of a sample temperatures using infrared thermography while an employee at the Dow Chemical Company.

Sharon was born on November 28, 1955, in Beaumont, TX. Her mother, Selna McDonald, was born in Sabine Pass, TX, graduated from high school, and went to Texas Southern University in Houston, TX. After college she attended business school, eventually becoming a high school secretary.[13] Sharon's father, William McDonald, a Silsbee, TX, native, attended Prairie View A&M College for two years and then returned home to work in the family grocery business.

Sharon has two sisters and a brother: Betty is eleven years older than Sharon; Willane, four years younger; and William, six years younger.[14] Sharon and her family went on numerous summer vacation trips, because her father wanted to expose his children to life outside of their small hometown. These summer trips inspired Sharon's love of travel. Her older sister and mother taught Sharon to read when she was four. Her parents always encouraged their children to attend college; it was never "if you attend college" in that family.

Sharon became interested in science at an early age, because her mother always served fresh fish on Fridays. Watching her mother dress the fish, Sharon was amazed at the internal organs and wondered how the fish could live in the water without air. Also, her grandfather died from complications from diabetes, and she wanted to understand the body and cure diseases.[15] With memories of the fish and with many questions in mind, she decided to become a scientist so she could find answers to these questions.

In junior high school, Sharon really enjoyed classes in both science and speech. In high school she was active in drama as well. Her math teacher, Mr. Ronny Nash, was an inspiration as was her high school mentor Mrs. Sherry Woodard. Mrs. Woodard, a registered nurse, inspired Sharon to pursue science and technology, telling her, "Hey, you've got a knack for this. You should explore it." Mrs. Woodard took Sharon's class on a field trip to the University of Texas MD Anderson Cancer Center[16] where they toured the cytology lab, which was studying breast cancer tissue. When Sharon saw the lab displays and understood the importance of its work, she began to think she might want to do something like that in the future.

Before Sharon graduated from Sisblee High School, the mantra at home was still "when you attend college." She graduated in 1974 and received the Edgar and Estelle O'Neill Award, from the high school which is scholarship given every four years to one white and one African American student. She applied to Texas A&M University in College Station and Baylor University in Waco, TX, visited both schools, and decided to go to Baylor because of its diverse student population. When she started college there she got off to a rocky start. Having taken a lot of courses in high school, she thought she could do the same thing in college and took too many courses for a freshman. She spent the rest of her time in college making up for that first year. In her first two years she took all the core courses: English, history, philosophy, and math. In her junior year, she started her major in chemistry and also took biology, which included anatomy and genetics, both of which she loved. She also did undergraduate research focused on the digestive system of diabetic Chinese hamsters, to determine if intestinal bacteria had any effect.[17] She also interned at the Baptist Hospital of Southeast Texas in the Clinical Laboratory Scientist program.[18] When Sharon graduated from Baylor in 1978, she had to decide what to do next. She thought about medical school, dental school, or working in industry. Having a love for research and knowing medical school or dental school would be cost prohibitive for her parents (there were two other children who needed to go to college), she decided to intern at a clinical Laboratory Science lab on the job market in 1979 she was hired at the VA Medical Center in Waco as a lab technician. During this same year, she received her certification as clinical laboratory scientist from the Veterans Administration at Baylor University. She worked there until 1981, performing liquid and gas chromatography procedures on toxicology specimens.

On February 14, 1981, Sharon married Ronald Barnes whom she had met in college. After graduation he became a high school teacher and coach, and, later, a high school principal. In 1982, Sharon gave birth to twins Ronald Barnes II and Amber Janel Barnes.

After her marriage, she left the VA hospital in Waco and moved to the Freeport-Lake Jackson Area where she was hired to work in the community hospital[19] laboratory, performing drug testing and blood banking as well as general bench chemistries for seven years. In 1987 she applied to work at Dow Chemical Company and kept calling their personnel department until they decided to interview her. She got the job, in their clinical lab, where she initiated their random drug testing program, establishing both the protocols for the chain of custody process and the facility where they would actually perform the tests. In 1991, she became the Special Chemistry Lab Supervisor for Dow and headed up the department because she was a specialist on the lab equipment.

The department medical director challenged Sharon to find a test to determine if collected samples were from a particular individual. It wasn't possible to simply stick a thermometer in the sample, because that would contaminate it. Sharon remembered that one of her friends had an infrared thermometer that he was using to check the pressure on steam lines. He let her borrow it, and Sharon took it back to her lab to figure out the temperature range of the specimens. Normal body temperature is 98.6°F and a voided urine specimen could be slightly above or below that. She worked on this problem until she solved it and the patent department decided to submit a patent application for the equipment and procedure. It took two years to get the patent, which listed Sharon and others as the inventors in 1991. The others on the patent were two engineers, two technicians, and the carpenter who designed the stand that the equipment sits on.

Sharon became a Dow-Texas Operations Inventor of the Year.[20] As she explained, "The above invention includes a process and apparatus for determining the temperature of a sample, such as urine, without contacting the sample itself. A portable device is used to carry the temperature measuring apparatus. The sample of urine is placed in a plastic container on an adjustable support and the temperature is measured by an infrared pyrometer."[21] This apparatus uses the same concept as infrared thermography. Because the patent actually belonged to the company, the company sold the license for about $100,000 a few years later. Sharon received a silver dollar as one of the inventors. The invention became the basis for the aural thermometers used to take a baby's temperature.

In 1993, Sharon served as Laboratory Director for the clinical labs at Dow; in 1994 she became the Clinical Lab Director in the Occupational Health Department. She also became a training specialist, which was the catalyst for her move to become one of Dow Chemical's Human Resources Business Partners. When she started her new position she had mentors in Dow to help her learn human resources. She learned her job by taking courses the company sent her to. But to understand the theory of human resources, she studied for and, in 2005, received a master's in business administration (MBA) in Human Resource Management from the University of Phoenix.[22]

Sharon had numerous mentors at Dow, including the vice president of human resources, Larry J. Washington, who was inspirational in her human resources (HR) career. Another mentor, Ms. Effie Durst, had very good insight in to the workings of HR, and on people in general, and was an excellent sounding board for her work. Sharon feels everyone should have at least one mentor, and perhaps two.

Her next position at Dow was Associate Director of Human Resources in the Performance Plastics Division. In this role, she supported a vice president and his team, assisting with strategic planning. As this was a global role, she interacted with employees and leaders in Europe, Latin America, China, Mexico, and Canada, and found it very interesting. In any given day, she could talk to or interact with someone across the globe. She also became Outstanding Scouter for Dow. Dow always encouraged young people to enter science and technology by taking them on field trips to observe open heart surgery at St. Luke's Hospital or to places like the hospital at MD Anderson.

Even as a black woman and a working mother in a corporation, Sharon has never experienced any problems with discrimination, neither toward herself nor her family. She is very proud of Dow in that respect.

Sharon ran for the city council of Lake Jackson, TX in 1996 and won, serving two two-year terms during which she enjoyed serving the people and listening to their problems. In 1999, when she was on the city council, she served as mayor pro tem for ten days while the major vacationed in Hawaii. She was able to run a council meeting and the city with no problems! In 2000 she campaigned for the position of mayor, but was not successful. Although she loved helping, Sharon did not like campaigning.

That same year, she was named a Distinguished Alumni from Baylor University, having been nominated by her daughter. And then-Governor George W. Bush appointed her to the Texas Medical Board; she serves on one of the District Review Committees, representing Harris, Brazoria, and Galveston counties. The Review Committee evaluates all the complaints that patients, insurance companies, or hospitals submit against physicians in Texas, provides oversight and advocacy for patients, and helps physicians modify unacceptable behaviors. The goal of the medical board is to protect the health and welfare of the citizens of Texas. As an appointee, Sharon serves at the pleasure of the governor and has served on the board for seventeen years. In 2004, Governor Rick Perry appointed her to the Texas Health and Human Services Council, on which she has served for four years.[23]

One of the Dow engineers who worked with her on the patent application was a member of NOBCChE, and Sharon also became a member. She got involved in the local section and was asked by then-secretary Dr. Marquita Qualls to run for the office of secretary of the national organization. So, in 2003, she became the National Secretary of NOBCChE.

Sharon considers getting her MBA and working with NOBCCHE the highlights of her career. She feels NOBCChE has made some great strides. The membership is growing; and the organization acts as a clearinghouse for technical entities, such as the NSF and National Institutes of Health (NIH). She receives great pleasure in having newer, younger people come up and say, "You know, I do want to go into science and technology. It's really interesting."

Her thoughts about the future? "We all need each other." All people need an education to get a good job. Some kids don't realize how tough and difficult life can be without an education. They shouldn't worry about getting a master's degree; just work on getting their high school diplomas first. Anyone can better their situation through science and technology. A career in science or technology can provide great job security. Young people have got to be willing to both reach out for and accept the help of older people, because we need each other. "Future scientists should always ask challenging questions and look for better ways to do just about anything. Always be willing to work hard, have faith, and trust yourself that you can make a positive difference," she says.[24]

What is Sharon's legacy? Always striving to do the best she could for her family and all the organizations with which she has been involved. She is proud to have been the first African American elected to city council in Lake Jackson. There are now two African American men on the council. She hopes that there will be no problems because of skin color if her grandsons decide in the future to run for elected office. She hopes that she leaves things better than she found them.

If she could have changed anything in her life, she would have studied more in college so as to make her job transitions easier. However, she is still satisfied, because everywhere she went she learned a little more and she could speak to other scientists from experience.

Sharon currently works for a fairly new company called Freeport LNG Development L. P. as their Senior HR Business Partner, having retired from Dow in 2014. She is passionate about helping this new company build a $14 billion project by recruiting and staffing the workforce with qualified and knowledgeable individuals. She says, "Freeport LNG is like a family to me, with one common goal, and I love it".

Her son, Ronnie, became an engineer because he was always interested in science. He would watch Mr. Wizard on TV and try to do the experiments he saw. Her daughter, Amber, once wanted to be a perfusionist working on heart-lung machines. But she changed her mind and, even though she loved science, decided to go into HR like her mother. Realizing she has a heart for students, she now teaches science and math in junior high school.

How Sharon would like to be remembered? "I'd like to be remembered as a loving, Christian woman who helped people wherever she could.

That'll be enough to be said. You know, all the accolades, and the degrees, and the appointments, are fine; but helping people and being a blessing to others is what counts. I just want people to say, 'Well, she was a Christian woman, she was loving, and she always wanted to help people.' That could be the best thing anybody could say about me."

2.4 Sherrie Pietranico-Cole

FIGURE 2.4. Sherrie Pietranico-Cole

Sherrie Pietranico-Cole (Fig. 2.4) is a former Senior Research Leader in Discovery Chemistry at Hoffmann-La Roche Inc. pharmaceutical company who now works in Drug Regulatory Affairs at Novartis Pharmaceuticals Corporation.

Sherrie's mother is Arline Pietranico and her father is Anthony Pietranico. Sherrie's African American mother was an operating room nurse at French Hospital Medical Center in New York City in the fifties and sixties. Her Italian father was a New York City correction officer and worked at Rikers Island. They married in 1964, a time when many still disapproved of white people marrying black people, and it was even illegal in some states, but not in New York. But, her parents did experience harassment. Sometimes, policemen passing in a patrol car, on seeing the young couple would pull the car over and beat her father up. Her father became a correction officer to protect his family—and this violence did end then.

Sherrie was born on July 22, 1965, in New York City. Her brother, Anthony Pietranico II, was born four years later. They grew up in in Harlem. Since her mother valued education, she worked hard to get her children full scholarships to private schools, believing they would get a better education in a private school where there were small classes. Therefore, Sherrie went to Brearley School, a prestigious all-girls private school in Manhattan, and Anthony went to Browning School. Since Sherrie was not happy at Brearley she transferred in middle school to the Nightingale-Bamford School for girls. When it was time for Sherrie to go to high school, there was a big change. Her mother inherited a house in West Milford, NJ, so her family moved and she went to high school there. The curriculum at Nightingale-Bamford school was so much further advanced than the curriculum at West Milford High School that Sherrie was allowed to skip eleventh grade on the condition she maintain her grades. She took biology, chemistry, and physics in high school and worked after school at Burger King to save money for college. She graduated from West Milford High School in 1982.

Paying for college was quite an issue. Sherrie knew her father would not help, because he believed that girls should not go to college—they should get married and stay at home to raise children. Sherrie also knew that her mother couldn't help, for she had stopped working as a nurse, at the request of her father, when Sherrie was four years old. Getting to the Burger King proved a challenge. At five miles away from her home, it was too far to walk; there was no public transportation; and her mother had never learned to drive. The solution was provided by a retired neighbor, Frank Lomax. He helped Sherrie reach her goals by driving her to and from Burger King every day after school, only charging her two dollars for the taxi service. Sherrie would often get home about eleven o'clock at night, stay up to study until two or three in the morning, and then get up and go to school. With the money she made at Burger King, Sherrie was able to pay for the first year of college.

Sherrie took the Scholastic Aptitude Test (SAT) one time and applied to colleges. Her father would not allow her to go to college outside of New Jersey and required that she come home every weekend. Sherrie felt very lucky to be accepted to Rutgers University where she attended Douglass College, the all-women college in New Brunswick, NJ, to study pre-law, because she wanted to be a lawyer. She took a diverse curriculum her first semester, including math, science, English, and history classes. For her science class she took General Chemistry with Professor Frank

J. Fornoff, loved it, and got an A. Professor Fornoff, retired from Princeton University, was rumored to be teaching for a dollar a year because he loved to teach. His class was held in a very large auditorium on the Busch Campus and Professor Fornoff assigned seats. During the class, he would call a student's name, ask a question, and expect a correct answer. Since he was such a remarkable teacher and taught chemistry so well, Sherrie decided to switch her major to chemistry. The class also included lab work, and the lab portion of the class was run by chemistry graduate students.

During the first year of college, Sherrie could not afford to live on campus and had an apartment on New Street in a poor section of New Brunswick. But she was happy there, because it reminded her of Harlem where she had spent her early childhood. She got a shock when she learned she had to pay for her own books, having thought that students were given books for classes as in high school. So, she got a job right away in the Douglas Student Center running the cash register, cleaning tables, and filling up the refrigerators. It was about two weeks before Sherrie received her first pay check, but even that wasn't enough to buy all the books she needed! Fortunately, her godmother sent her the rest of the money to buy her books.

Sherrie's best friend in college was Sonya Thompson. Sonya's father was an executive at International Business Machine Corporation (IBM). He was impressed with how hard Sherrie worked to pay for college, and the summer after freshman year, he helped her get an internship with IBM working in the parts center. She was the only woman there and her job was simply to fill orders for parts, but she made enough money that summer to pay for her sophomore year of college and to live on campus.

During her sophomore year, Sherrie saw an ad for a mentoring program sponsored by the National Coalition of 100 Black Women Inc. Designed to teach young women to write a resume and dress for an interview and to introduce students to mentors working in the field they wanted to pursue, the program required students to apply to be accepted. Sherrie applied, was accepted, and invited to visit Merck by Ms. Susan Jenkins, the first African American woman to work as a chemist at Merck. During the visit Sherrie met her mentor, Jeannette Brown, who was working there a chemist. Sherrie learned of and applied for a summer internship at Merck for the summer after her sophomore year. She was offered an internship in Medicinal Chemistry and worked with Dr. Philippe Durette, who remains in contact with her still. He taught her how to set up reactions, purify compounds, take ^{1}H NMR (proton nuclear magnetic resonance) spectrums of the compounds, and interpret the NMR spectrums to determine if they had synthesized the right compound. If not, Dr. Durette would propose another reaction to get the correct compound. Sherrie asked him how he knew all this, and he told her it was from getting a PhD in Chemistry, and if she wanted to know more, she should get her PhD. She said she could not afford it. He told her students attend for free in PhD programs in chemistry. He also encouraged her to do undergraduate research at Rutgers when she returned to school. When she got back to Rutgers, she contacted Professor Spencer Knapp about research. He accepted her into his group and assigned her to his graduate student Paivi Kukkola, who became another mentor. Paivi showed Sherrie additional ways to synthesize compounds. Sherrie would go to class, study, and work in the lab. She worked on a novel proposal to synthesize cis-vicinal amino alcohols, which in 1987, became her first publication in *Tetrahedron Letters*.

In her junior summer, Sherrie returned to Merck to intern in process chemistry, working with Dr. James M. McNamara, who also became a mentor. Process chemistry has a different emphasis than medicinal chemistry. Medicinal chemistry involves making different analogs and studying structure–activity relationships to decide whether a change in the compound helped the biological activity. Process chemistry involves optimizing the synthesis of one compound. Sherrie worked on optimizing the synthesis of an HMG CoA reductase (3-hydroxy-3-methyl-glutary-coenzyme A reductase) inhibitor—a drug to lower cholesterol that Merck wanted to bring to market. Dr. Seiji Shinkai, Head of Process Chemistry, invited Sherrie to his office to give a "chalk talk" on her progress, with no notes. Since a chalk talk meant that she had to speak about her work from memory she found it challenging but was glad of the experience because it helped her prepare for giving talks using a black board in graduate school.

Her participation in those summer internships at Merck and in undergraduate research with Professor Spencer helped Sherrie get accepted to many PhD chemistry programs. Her mentors from Merck encouraged her to go to the University of Pennsylvania (Penn) to get a PhD in organic chemistry with Professor Kyriacos C. Nicolaou. Sherrie was the first African American woman to be accepted into the PhD chemistry program at Penn. That year Professor Nicolaou announced that students who received an A in his classes, Organic Synthesis I and II, would be accepted into his group. Two A-students, Sherrie and David Nubile, accepted offers to join the group. Sherrie became the fourth woman to be accepted into this group. The third woman was Patricia (Patty) Somers who was getting ready to graduate and also mentored Sherrie. They shared a desk and bench space because there was no free space when Sherrie arrived.

Shortly after Sherrie joined Professor Nikolaou's group, Dr. Ralph F. Hirschmann, the Head of Research and Development at Merck, retired and came to Penn as a professor. Professor Hirschmann shared graduate students with the other professors. Since Sherrie had interned at Merck and was in awe of Professor Hirschman, she lobbied for and earned the honor to be his first graduate student. She says she was incredibly fortunate to learn from Professor Nicolaou, one of the world's greatest synthetic chemists, and Professor Ralph Hirschmann, one of the world's greatest medicinal chemists. She completed a medicinal chemistry project with the two men, discovering peptidomimetics of somatostatin, a fourteen-amino acid peptide, which is present in humans and regulates glucagon and growth hormone. It can inhibit the production of glucagon and help patients with diabetes and inhibit the production of growth hormone and help patients with acromegaly. When Professor Hirschmann was at Merck, his team discovered that just four of the fourteen amino acids of somatostatin were responsible for its biological activity, that they form a β-turn conformation, and that the side chains of these four amino acids, not their amide bonds, were important for the biological activity. Normally with peptides, amide bonds are critical for biological activity but scientists at Merck synthesized a peptide analog of somatostatin in which the amide bond was reversed and it had similar biological activity to somatostatin, demonstrating that the amide bonds were not critical for biological activity, but the side chains were. Professor Hirschmann wanted to make peptidomimetics of these four amino acids, which meant discovering a scaffold which would position the four side chains in the same conformation as in somatostatin. At that time, Penn had just received a Silicon Graphics Computer for molecular modeling, which the team used to design the peptidomimetics of somatostatin. Sherrie and others Penn students helped synthesize them and were delighted to discover the first peptidomimetic agonist of somatostatin. Professor Hirschmann had maintained a close relationship with Merck, and he sent the compounds for further testing there, where scientists discovered that the compounds were also potent substance P antagonists, possibly useful for the treatment of pain.

Sherrie became the first African American woman to graduate with a PhD in chemistry from Penn. When Sherrie arrived there, Bruce Harris, the fourth black man to get a PhD in Chemistry from Penn, was in his last year of the program. Sherrie went to the chemistry office and asked them why there was so little diversity in the program. She thinks that meeting made a difference, because after her year, one African American student a year was accepted into the program.

During Sherrie's first year in the program, she was given a cumulative exam on the Dr. Martin Luther King Jr. holiday, as was traditional at Penn. Instead of taking the exam, Sherrie wrote an essay on Dr. King in the exam booklet and expressed disappointment in a tradition that set an exam on such an important holiday. The following year Penn changed the test date.

It was a requirement of the PhD chemistry program that students be TAs. In her first year, Sherrie was a TA for general chemistry and general chemistry lab; for the remaining years, she was a TA for organic chemistry and organic chemistry lab. As a TA, she supervised undergraduates during the lab classes, went over the problems assigned for homework, prepared students for their exams during recitation, and graded lab books and exams. For this work she received a stipend of $1,000 per month. She was really happy because it was, enough money to pay her rent and buy her food.

Sherrie lived in the Graduate Towers at Penn, where the apartments had two bedrooms and a kitchen—she had her own bedroom; her roommate was from Morocco. After two years, she moved off campus. After her first apartment was broken into, she moved to a building with a doorman.

The graduate students and post doctorate students (post docs) in Professor Nikolaou's group were very diverse and came from all over the world, many from Greece and Japan. Sherrie was the only African American in his group and the only woman for a year after Patty left. Even though she was a different gender and race, she felt welcomed by the other scientists in the group. Sherrie's believes scientists are special. Scientists generally see each other as scientists and generally don't treat people differently based on gender, race, or ethnicity. She was also able to learn so much outside of chemistry by learning about the culture of the different students in Professor Nicolaou's group—the Japanese post docs introduced her to sushi. It was a fantastic experience.

In her last year at Penn, Sherrie started interviewing for jobs and received many offers. She remembers Professor Smith coming to Professor Nicolaou and Professor Hirschmann and telling them she should not interview anymore because she had enough offers. He thought it was time for her to pick a job. Sherrie loved interviewing because she was usually taken to a really nice dinner, something she could not afford to buy herself as a graduate student. She got an offer from Johnson & Johnson that would have meant working with Dr. Cynthia A. Maryanoff, a powerful, brilliant woman, which she wanted. But she was torn because the work was in process chemistry and Sherrie really liked medicinal chemistry. So, she decided to accept a job at Hoffmann-La Roche Inc. (Roche), a pharmaceutical company in Nutley, NJ. She was attracted to this company because of its Institute of Molecular Biology and its staff that included some very well-known scientists. She also wanted to be close to New York City where her boyfriend and future husband, Dr. Brian A. Cole, had an orthopedic residency at St. Vincent's Medical Center.

Sherrie started working at Roche before she defended her thesis. So, she wrote her thesis at night while she was working, then went back to Penn to defend it. She held a forty-five-minute seminar on her research for the Chemistry Department, then took her place in Professor Edward R. Thornton's office to face the professors on her research committee. They all asked her questions and she was able to answer all but one. Professor Hirschman was angry that she had been asked that question and said, "If she could answer that question, she'd deserve a Nobel Prize in Chemistry." He said even he could not answer that question. Sherrie passed her defense and graduated in May 1992, becoming the first African American woman to graduate from Penn with a PhD in Organic Chemistry.

This was the same year the first woman African American astronaut, Dr. Mae C. Jemison, flew to the moon and Hillary Clinton spoke at Penn's graduation. Because of these women's accomplishments, Sherrie thought that women were really making progress, and expected even more progress to follow in the 1990s. She even wondered when the first African American woman would be elected president. But, Sherrie was upset about what was happening to black South Africans under apartheid and dedicated her thesis to them in the hope that one day they would have freedom to reach their full potential.

Sherrie knew she had grown up poor, but in spite of that she still had freedom, unlike black South Africans living under apartheid. That freedom gave her the ability to change her life. When Sherrie was growing up, there was sometimes no heat in the apartment, and her mother had to turn on the oven to keep one room warm, where they all slept. Sometimes there was no hot water, so they would boil water on the stove to take a bath. There were roaches everywhere. Some nights they would have to lie on the floor to keep from getting killed, because people would shoot guns outside. In the morning, she would see bullet holes in the building. In fact, Jesse Gray, a well-known Harlem drug dealer lived in her building, and she was very much aware of the drug dealing going on around her. Despite these conditions, she had access to some very prestigious schools. Sherrie knew that with good grades and a good education she could have a better life and chose that path even as a young child. The one thing she hopes that anyone would learn from her story is that with opportunities, hard work, perseverance, dedication, and good grades, you can change your life.

Sherrie started as a Senior Scientist at Hoffman LaRoche (Roche) on March 2, 1992. Her first project was for the treatment of asthma. The goal was to discover LTD4 antagonists that fixed some of the liabilities of the LTD4 antagonist then in development. Sherrie threw herself into the research and was often in the lab until midnight. She worked with a brilliant medicinal chemist, Dr. Armin Walser, who would do reactions in a test tube to see if they would work. Under Armin's leadership, Sherrie's team discovered five compounds with properties that were superior to the compound in the Merck drug. But, Roche decided not to develop the drugs.

Sherrie also worked with another brilliant medicinal chemist at Roche, Dr. Louise Foley. When Dr. Foley showed up at the Massachusetts Institute of Technology (MIT) to get her PhD in chemistry, the professor she was supposed to work under would not accept her into his group because he had thought she was a man, Louis, when she applied. Another professor, Nobel Laurate Professor George Buchi took her into his group. Sherrie, Louise, and others worked on PEPCK inhibitors for Type II Diabetes and their work is published in *Bioorganic & Medicinal Chemistry Letters*.

Sherrie worked very closely with her assistants. John R. Vermeulen, Sherrie's first assistant, worked with her for fourteen years until he retired; Qiang (Alan) Zhang, worked with her for six years; and Bill May and Bill Zally worked with her for a few other years.

Sherrie was working to help develop and market a drug in partnership with another company Genetech Inc. She was able to work well with the teams in order to get the drug to market. It was because of this skill she was able to work with other teams as a leader in developing and marketing new drugs. She went on to lead many projects for Roche, two of which made it into the development group to become as drugs. Since most projects last about three to five years, and may be stopped if the compound could not be made properly or could not become a viable drug, she was very proud about those becoming development candidates.

The first compound that made it into development was an HSD1 inhibitor for Type 2 Diabetes, a joint project between chemistry teams in Nutley and Basel, Switzerland. Sherrie's team formulated chemistry strategies which improved potency and drug properties, including pharmacokinetics and safety. It was very exciting for her team when this clinical lead drug was accepted into the Metabolic Diseases portfolio. Compounds have to be tested before they can be declared as viable drug candidates.

The second compound being developed was a thyroid hormone agonist for the treatment of high cholesterol, metabolic disorders, and non-alcoholic steatohepatitis (NASH) (also referred to as Non-Alcoholic Fatty Liver Disease), designed to be thyroid hormone receptor beta selective. Thyroid hormone regulates energy metabolism and affects fat distribution, body weight, and metabolic disorders. It also has deleterious effects on the heart. The beneficial effects of thyroid hormone on lipid levels are primarily due to its action at the thyroid hormone β (THR-β) receptors in the liver, while adverse effects, including cardiac effects, are mediated by the thyroid hormone α (THR-α) receptors in the heart. The goal was to design a selective thyroid hormone β (THR-β) receptor agonist.

It was very challenging to design a drug to selectively work in the liver. Sherrie's team worked very closely with Dr. Irwin Klein, Chief of Endocrinology at North Shore University Hospital and an expert on thyroid hormone, to screen for compounds that would not affect the heart. He tested the team's compounds in thyroidectomized rats to determine their effect on myosin heavy chain (MHC) gene expression in the heart. Thyroid hormone regulates transcription of proteins critical to heart function, like MHC. When the thyroid is removed, the heart only has beta myosin heavy chain (β-MHC hnRNA) gene expression. If thyroid hormone is given to these thyroidectomized rats, the gene expression in the heart changes to alpha myosin heavy chain (β-MHC hnRNA) gene expression. Dr. Klein tested the Roche compounds to find compounds that wouldn't affect β-MHC hnRNA gene expression the heart. If the gene expression did not change in the heart, then the compounds were not affecting the heart. That's how they found the clinical candidate. The team was extremely excited when one compound, MGL-3196, was discovered which had little to no effect on gene expression in the heart at very high doses but had very beneficial effects in the liver. When MGL-3196 was administered to DIO (diet induced obese) mice, it reduced fat in the liver to normal levels, reduced cholesterol, reduced body weight, had no significant effect on food intake, and increased energy expenditure at doses that showed no impact on the central thyroid axis and no significant effects on the heart or bone.

The team that did this work was a high-performing team constantly brainstorming together. They accomplished their goal, discovering a very safe compound that was a selective thyroid hormone β (THR-β) receptor agonist. Roche takes a conservative approach to safety and Sherrie was nicknamed the "Queen of Safety" because she was very focused on the safety of her compounds. The selective thyroid hormone beta receptor agonist made it through Phase 1 studies in healthy volunteers and behaved precisely as predicted. MGL-3196 exhibited an excellent safety profile and an acceptable pharmacokinetic profile and decreased LDL cholesterol (LDL-C) and triglycerides (TG) at once with daily oral doses of 50 mg or higher when given for two weeks. These results were published in *Atherosclerosis* MGL-3196 is currently being tested in Phase 2 studies. Sherrie hopes it will transform the treatment of Type 2 Diabetes and NASH.

In 2008, because of her contributions toward discovering medicines at Roche, Sherrie was awarded the Tribute to Women in Industry (TWIN) Award,[25] which is given to women who have made noteworthy contributions to their industries. Sherrie rose to the highest level on the research ladder at Roche and continued to lead projects until 2011. She is listed as inventor on sixteen patents and she has won three inventor patent awards.

After nineteen years of discovering drugs at Roche, Sherrie decided to change careers and take a job in Drug Regulatory Affairs in 2011, at a time when most employees at the Roche Nutley site expected layoffs. Roche had bought Genentech and it was becoming clear that huge changes were in the offing. No one expected the whole site to close, though. Sherrie moved to Drug Regulatory Affairs because she knew that there would be many opportunities to get a job in that field in New Jersey if she was laid off. Her husband was established in New Jersey by then, so she needed to stay and find a job there. In her new job Sherrie worked on Tamiflu, a drug to prevent and treat flu. She was part of a team that filed a supplemental New Drug Application (NDA) for the treatment of children under one-year-old with the flu. She also worked on Pegasys, a drug to treat Hepatitis C, and she was the regulatory lead for some programs early in development. Sherrie worked in Drug Regulatory Affairs for about one year before Roche closed the Nutley site and laid her off in October 2012. She decided that she had made a good decision, even though she had not wanted to leave chemistry, as she was offered a job three weeks later at Novartis Pharmaceuticals Corporation in East Hanover, NJ, in Early Development Drug Regulatory Affairs. She works with the Novartis Institute of Biomedical Research (NIBR) teams as soon as a clinical trial candidate is identified.

The regulatory leader on a project is the liaison between the company and the health authorities. In the United States, this authority is the Food and Drug Administration (FDA) and in Europe, it is the European Medicines Agency (EMA). Additionally, each country in Europe has its own national health authority and there are health authorities in the rest of the world. The drug regulatory affairs leader develops a regulatory strategy with the teams in each country and works with them to get approval for clinical studies in the countries where the study will be done. Sherrie likes working in drug regulatory affairs because she stays close to the science. At Novartis, she works with the early development teams and advances compounds into Phase 1 clinical studies through Phase 2 proof-of-concept studies. She leads the preparation and filing of the Investigational New Drug Applications (INDs) with the FDA to get approval to conduct the First-in-Human (FIH) Phase 1 studies. Sometimes the FIH study is run in Europe or in other parts of the world, and Sherrie helps the teams prepare and achieve approval of a Clinical Trial Application (CTA). INDs and CTAs contain documents which summarize the pharmacology, pharmacokinetics, and toxicology and aim to show health authorities the benefits and risks of a compound. After the FIH study is completed, Sherrie helps achieve approval of the Phase 2 Proof-of-Concept (POC) studies. In the past five years, Sherrie has helped her teams achieve approval of CTAs in Brazil, Belgium, Canada, Germany, Netherlands, United Kingdom, South Korea, and Taiwan.

Another important aspect of Sherrie's job is formulating the regulatory strategy with her teams. She gathers regulatory intelligence to guide the teams. Her job is to understand the Code of Federal Regulations, interpret those regulations for the team, and generally ensure the teams know both what the regulations say and options to apply them. In addition, she ensures the teams are aware of FDA and EMA development guidelines, which are published by the health authorities, outlining the regulatory strategy, along with the regulatory risks and risk mitigation plans with the teams.

The teams have experts to cover every aspect of drug development. The members of a team typically include a clinician, a Clinical Trial Leader (CTL), a statistician, a safety expert, a drug metabolism pharmacokinetic expert, the preclinical biology leader, and the preclinical chemistry project leaders. There usually are also a microbiologist (or biochemist) that runs the in vitro assays and an in vivo pharmacologist that performs studies in animal models, as well as a biomarker expert. The project manager is in charge of the overall strategy, the timelines, and the budget. There is also a patent lawyer, a technical expert to oversee drug manufacturing and supply, a formulations expert, a Drug Supply Manager, and a Chemistry Manufacturing and Controls (CMC) Regulatory Leader. There could also be a Regulatory Device Expert. They meet together to formulate the strategy to meet their goals and meet with management for support of the proposed development strategy.

Sherrie prepares the team for the health authority meetings she requests. Sometimes the teams need to get agreement with health authorities on their plans. For example, agreement may be needed on study design, the endpoints for the study, the safety monitoring plan, the required toxicology, or the registration studies. Overall, there is a lot of strategy involved with being the regulatory leader, which makes it a very challenging but rewarding job. Sherrie thinks that Drug Regulatory Affairs is a great alternative career for chemists.

Sherrie spends her time outside of work with her family: her husband, Dr. Brian Cole, daughter Jasmine, and son Brian Jr. She met her husband at Penn at a party organized by the Black Graduate and Professional Student Association (BGAPSA) to bring black students together from all the different schools at Penn. Brian was in the medical school at Penn while Sherrie was working on her PhD.

Sherrie was fortunate that her mom cared for her children while she worked. Her mother had learned to drive by then and would pick the children up from school and stay with them until Sherrie came home from work. Her mom did that every day until she became ill with terminal cancer. When Sherrie came home, she'd cook dinner, and everyone, her family and her parents, would eat together before her parents would go home.

Sherrie is involved in community service and values giving back. She especially enjoys inspiring children with an interest in science. Roche had a tutoring program for third grade students from Paterson, NJ, who were mostly African American and Hispanic. Sherrie set up a science program for the students which was a big highlight. She would tell them about African American scientists, and the students performed hands-on experiments in their honor. The students also got to develop their own product, a bubble bath, something they could take home and be proud of. She has conducted science programs at the Jersey Explorer Children's Museum in East Orange, NJ, and organized science fairs for elementary schools in South Plainfield and Montclair, NJ. Sherrie volunteered at an ACS Science Day at Liberty Science Center in Jersey City, NJ, helping design the experiments and spending the day there. She also helped design and really enjoyed volunteering for Science Olympiad's Experimental Design competition for middle and high school students. Sherrie and colleagues from Discovery Chemistry at Roche would go to New Jersey Institute of Technology (NJIT) to set up the science materials, oversee the competition, judge their designs, and decide on the winners. Sherrie also allowed a science teacher from Newark, NJ, to come spend a summer in her lab at Roche as part of a program to help science teachers gain experience that would improve the quality of science education. She now does science programs at Novartis with the students in their mentoring program.

While Sherrie was at Roche she also mentored INROADS and Project SEED[26] students by allowing them to do summer internships with her and gain experience in the lab. INROADS is a program for underprivileged minorities in college. The undergraduates attend trainings on weekends learning how to write a resume, conduct an interview, and seek summer internships. One of Sherrie's students, Barbara Rodriquez, now has a PhD in Pharmaceutics. Barbara was a Project SEED student and did an internship with Sherrie as a high school student and Sherrie helped her get accepted into the INROAD program at Merck, which also helped Barbara pay for college, because Merck gives the students ten thousand dollars toward their college expenses. Victor, another student, was also an INROADS college student. Sherrie won the INROADS Supervisor of the Year award one year and the INROADS Spirit Award another year. Being a mentor to minority and underprivileged students is very important to Sherrie. She was mentored by so many people who encouraged her and contributed to her success, she feels that she must give back. She also finds this very rewarding and has decided to teach when she retires.

Sherrie belongs to NOBCChE and thinks the excellent career fair it organizes is very important, really exposing African American scientists to great opportunities. She was able to recruit three chemists for Roche at NOBCChE national meetings. She also belongs to the Association for Women in Science (AWIS), the ACS, Drug Information Association (DIA), and Regulatory Affairs Professional Society (RAPS). Sherrie was inducted into the Chemistry Honors Society, Phi Lambda Upsilon, in 1988 while she was at Penn.

When asked about her perception of the chemistry profession now, Sherrie says she is sad. Currently there are far fewer jobs in chemistry. She was torn, because she thought that the future for research chemist was dim, when she saw President Obama pushing STEM and she wonders if the job prospects will change in the future. Many chemistry jobs have been outsourced to China and India. When Roche closed the Nutley site, some of her colleagues had to go to China and India to get a job in chemistry. If Sherrie were an ACS career advisor she would advise students to get a Doctor of Pharmacy (PharmD) degree and pursue a career in drug regulatory affairs where there are more job opportunities.

What advice would she give young minority students who want to study science? Study hard and get good grades. That could change your life. If the students attend public schools, challenge yourselves. Take honors and Advanced Placement (AP) classes.

Sherrie said if she were an ACS Chemistry Ambassador and could talk to a congressperson, she would warn them that most of the pharmaceutical companies have outsourced the jobs in chemistry. She thinks that they are misleading US children by encouraging them to go into STEM without bringing science jobs back to the United States.[27]

Notes

1. See her story in Jeannette E. Brown, *African American Women Chemists* (Oxford University Press, NY, 2012), chapter 5, 46.
2. Tiananmen Square incident, also called June Fourth or 6/4, was a series of protests and demonstrations in China in the spring of 1989 that culminated on the night of June 3–4 with a government crackdown on demonstrators in Tiananmen Square in Beijing
3. See her story in Brown, *African American Women Chemists*, 92.
4. Dorothy Phillips, interview by Jeannette Brown, April 5, 2013, Oral History Transcript, Science History Institute, Philadelphia, PA.
5. A solid substance that has a conductivity between that of an insulator and that of most metals, either due to the addition of an impurity or because of temperature effects. Devices made of semiconductors, notably silicon, are essential components of most electronic circuits. English Oxford Living Dictionary

Definition of semiconductor http://www.oxforddictionaries.com/us/definition/american_english/semiconductor accessed May 9, 2018.

6. In 2001, the American Society for Testing and Materials was renamed ASTM International. For more information about the organization, see http://www.astm.org.

7. Gas chromatography is an analytical machine which breaks down the injected material into gas and analyzes the components of the material.

8. Pittcon is the leading annual conference and exposition for laboratory science, new technology, and scientific research. Its name came from Pittsburg, PA, where the first conference was held.

9. See Selected Publications.

10. As a part of the Victors program they would: 1. Pull and research the files in the court. 2. Fill out the form that is used by the visitor with date and address of the ward and the guardian to make it easier for the visitor to contact and visit and 3. An auditor would come and check the financial situation of the ward.

11. For a discussion about what the ACS committees do see https://www.acs.org/content/acs/en/membership-and-networks/td/manuals/committees.html.

12. Cherlynlavaughn Bradley, interview by Jeannette Brown, 6 April 2013, Oral History Transcript, Science History Institute, Philadelphia, PA.

13. Otha Richard Sullivan, "Sharon Barnes," in *Black Stars: African American Women Scientists and Inventors* (New York: John Wiley & Son, Inc., 2002), 108.

14. Living History Questionnaire written by Jeannette Brown and filled out and submitted to Jeannette Brown by Sharon Barnes.

15. Ibid.

16. The University of Texas MD Anderson Cancer Center is one of the world's most respected centers devoted exclusively to cancer patient care, research, education and prevention.

17. Ibid., iii.

18. Sullivan, "Sharon Barnes," 110.

19. Now called Brazos Port Regional Health Care Center.

20. Mitchell C. Brown, Sharon Barnes "The Faces of African Americans in the Sciences," [https://webfiles.uci.edu/mcbrown/display/barnes.html. A team of five (two African Americans) were assigned U.S. Patent #4,988,211 for an application in Infra-Red Thermography.

21. Sharon J. Barnes, MBA/HRM https://www.linkedin.com/in/sharonmcdonald-barnes/ accessed May 9, 2018.

22. Living History Questionnaire written by Jeannette Brown and submitted by Sharon Barnes.

23. Sharon Barnes The Science Makers http://www.thehistorymakers.com/biography/sharon-barnes accessed May 9, 2018

24. Sullivan, "Sharon Barnes," 110.

25. http://www.prnewswire.com/news-releases/two-roche-women-honored-as-2008-tribute-to-women-and-industry-awardees-57089532.html.

26. Project SEED. was established in 1968 to help economically disadvantaged high school students expand their education and career outlook.

27. Sherrie Pietranico-Cole, interview by Jeannette Brown, April 25, 2016, Oral History Transcript, Science History Institute, Philadelphia, PA.

3

Chemists Who Work in Academia

3.1 Etta C. Gravely

FIGURE 3.1. Etta Gravely

Etta Gravely (Fig. 3.1) is a retired professor of chemistry and former head of the Department of Chemistry at North Carolina A&T State University at Greensboro (North Carolina A&T).

Etta was born on August 30, 1939, in Alamance County, NC. Now the town of Green Level, it was then a rural community near Burlington. Most of the people there farmed, raising tobacco. Everyone had private gardens and Etta's grandmother canned their food. The area where she went to school is still very rural; the school building is now the town hall.

Etta's mother was Kate Lee McBroom and her father Rufus Leith. Her mother, a homemaker, did general house cleaning for families. Her father had a high school degree, had served in the army during World War II, and worked as an orderly in a hospital. Etta is the only child of her mother, but her father had a son named Frederick Leith. Her brother went to Graham Central high school and upon graduation went into the army and subsequently died.

Etta did not go to kindergarten because there was none. She started school in the first grade in a four-room school that had classes for grades one and two, three and four, five and six, and seven and eight. The principal was Mrs. Mary Holne, and there were three other teachers, each teaching two grades. Since Etta loved to read and liked to do school work, she skipped fourth grade and went on to fifth grade: fourth and third grade were taught in the same room, and when she completed her third-grade work she would do fourth-grade work. Her teachers probably had bachelor's or master's degrees in their subjects. Both Etta's school and community were segregated; she went to school in 1945, before the Brown vs. Board of Education act, which was Supreme Court decision.

When Etta graduated from the country school, she was bused to Pleasant Grove High School—for African American students, five miles from the high school for white students. The school taught grades one through twelve; the curriculum was the usual reading, writing, and arithmetic. They had a student council, class officers, a class song, and poem, and an elementary school girl and boy as mascots for the senior class. The high school had athletics and organizations in which students could participate. They had sports teams for boys and girls and cheerleaders for all the teams. Etta was captain of the cheerleaders and served as vice president of the New Homemakers of America (NHA) club. The thirteen school bus drivers were all students; their former principal's daughter, Gladys Marie Morris, and Etta were the only female drivers. The students published a yearbook called the *Panthers* and were allowed to leave school early to visit various businesses in Burlington to solicit ads to fund the publication. They prepared speeches and forms for the businesses to complete and probably raised enough money from the ads and patrons to pay for the books. They also sold yearbooks.

Their school books were probably books that had been used and discarded by the white schools.

Etta graduated from high school in 1956 and went to Howard University in Washington, DC. She decided to major in chemistry because of her experience with high school chemistry: her teacher Mr. Rivera G. Mitchell made things so interesting that everything just came alive for her. He had a bachelor's in science from North Carolina A&T and a master's in science from North Carolina College in Durham. In high school, although they may not have had all the equipment that they needed, they had enough to satisfy their goals at the time. They had enough for the students to excel on standardized exams and receive scholarships. The students took national standardized exams, including the SAT, at North Carolina Central University in Durham, because that school gave scholarships.

Etta had a great time at Howard and worked in the chemistry department. Dr. Lloyd Noel Ferguson Sr. was chairman of the chemistry department at the time. Other professors were Dr. Kelso Morris and Dr. Maddie Daniel Taylor. Ms. Jessup was the librarian, and Etta sometimes worked in the library. She was able to take undergraduate courses with some of the graduate students, as many of the graduate and undergraduate courses were combined, though the graduate students were required to do more in the course than the undergraduates. She especially remembers taking instrumental analysis with them. She joined AKA in 1957 and found that social experience was great. She did not do undergraduate research that was required of the graduate students.

All of her professors were very accomplished. Dr. Joseph B. Morris taught analytical chemistry and instrumental analysis. Dr. Robert Percy Barnes, the first black student to receive a PhD in organic chemistry from Harvard University, taught the advanced organic course. Dr. W. Peter Hambright taught inorganic chemistry. Dr. Lloyd Nelson Ferguson, in addition to being chair of the department, taught the first semester of organic chemistry, using his own textbook. Etta has subsequently interacted with all her old professors as an employee at North Carolina A&T.

She had met her future husband, Clinton Eugene Gravely, before she went to college, but they didn't date until they were in college. She reconnected with him at Howard University, reminding him they were both from North Carolina. On campus the Carolinians stayed together and joined in activities together. In addition, her roommate was her future husband's sister. Etta and Clinton married in 1960 after her graduation and moved back to Greensboro, where Clinton joined the architectural firm of Edward Lowenstein and Robert Atkinson.

Etta looked for a job in Greensboro, but she had little success—the first job she found was too far out of town for a daily commute. The employment agency was segregated; whites on one side, blacks on the other. When she went to the white agency they asked if she were Indian. When she said "no," they sent her to the black side where there were no jobs that she wanted. Some of the companies she went to only wanted males because they needed people with the strength to move huge vats around.

Since she could not find a job and seemed to have no other option, Etta decided to go back to school to get a teaching license. She had not taken teaching courses at Howard because she had not planned to teach; she had hoped to find a profession in chemistry. But, now she enrolled at North Carolina A&T, and when she finished the teaching courses, she applied for a job at Dudley High School, the only African American high school in Greensboro. Schools were not yet integrated.[1]

Dr. John Tarpley was the principal at Dudley. Etta was told that he was only hiring locals who lived in Greensboro and who had a master's degree. At the time she was living in Reidsville and hadn't thought about a masters. But her teachers at North Carolina A&T called Dr. Tarpley, told him she was coming for an interview, and highly recommended her. At the interview, when he discovered she had a degree from Howard he considered her. A few days after her interview, he called her, reported he had checked her references, and told her she had the job.

Etta taught in the high school from 1962 to 1966. Then she went back to North Carolina A&T to get her master's degree in education with a concentration on chemistry. She went back to Dudley to teach and finished her research for the master's of science in chemistry. Then back to North Carolina A&T to complete more research. She has two master's degrees, one in education with a concentration on chemistry and one in chemistry.

Etta liked teaching at Dudley; she thought it was fun. She encouraged some of her students to major in chemistry and has followed their careers. Some of them went to Howard and majored in chemistry; one went to Bennett College in Greensboro; and some went to North Carolina A&T. One, Rose Shaffer, got a bachelor's degree in chemistry and went on to law school at North Carolina Central University before going to Boston to work as a patent attorney, finally becoming a patent attorney at Ciba-Geigy Corporation.

Etta taught at Dudley High school until her third child, Thyrsa, was born in November of 1969 when she became a stay-at-home mom.

But, in 1970 she decided to apply for a PhD program in textiles at the University of North Carolina at Greensboro (UNCG). As UNCG had neither the equipment nor the certified faculty for a full-fledged degree program, it partnered with North Carolina State University in Raleigh. Therefore, Etta would have to attend courses at UNCG for the first two years and then go to North Carolina State University in Raleigh for the last two years. She was encouraged to continue taking courses assuming she would be formally admitted to the program. But there was a lack of action on her application, which she thinks was due to racial discrimination. The person who was supposed to review her application was on sabbatical and nothing happened until their return. Etta was retroactively admitted to the program after she had been attending classes for a year. By that time, she had decided that this was not going to work for her, due to the discrimination she experienced, so she decided to do something else.

Etta found out that Sarah Sullivan, wife of Dr. Sullivan (chair of chemistry at North Carolina A&T) was teaching some of the physical science classes at North Carolina A&T. Sarah Sullivan was asked to take a job at Grimsley High School teaching biology, which she wanted to take. However, they needed somebody to teach physical chemistry and physical science at North Carolina A&T. So, Sarah called and asked if Etta would take her place teaching physical science at North Carolina A&T. Etta agreed to take the part-time job, with only a master's degree. This was in 1972. Etta taught there for five or six years. Then the North Carolina Board of Governors offered an award to allow teachers to complete a terminal degree (the highest degree given in a particular field), which would include salary for a year of study.[2] Etta applied to the program in 1978 but was not selected. When she asked why, she was told they were looking people within one year of completing their degree, as they needed more people with terminal degrees in the school system.

About that time, Etta learned about the 1978–1979 Doctoral Study Assignment Program that was available at twelve of the sixteen institutions of higher learning in North Carolina. Priority was to be given to faculty members at traditionally black institutions because of the lower proportion of faculty having doctorates at these institutions and the national shortage of African Americans holding a terminal degree in their profession. The recipients of these awards were essentially given sabbaticals for a year. Etta applied and was accepted and so received the salary and benefits of her budgeted position and was employed by North Carolina A&T even though she was not there. At the end of her first year, Etta applied again, was accepted, and went to school full time from 1980–1981, at UNCG. With the help of these grants, Etta received an EdD, a Doctorate of Education in curriculum and instruction. Her doctoral thesis ("Alternative School in North Carolina 1977–1981") was about such schools as the North West School of the Arts in Charlotte and the Health Sciences Academy in Dudley, magnet high schools for students with a particular interest who would go on to pursue those interests, such as art or health science, at college. These students entered college with an interest in the things they had learned in high school.

Etta's doctorate was used to good effect in several ways. Etta chaired the curriculum committee in the chemistry department and served on both the senate and faculty assembly for the state of North Carolina. The University of North Carolina (UNC) Faculty Assembly was established in 1972 at the request of then UNC President William Friday. The assembly serves the president, UNC General Administration, the Board of Governors, the NC General Assembly, and campus faculty and administrative bodies, as a faculty advisory body on system-wide issues. In addition to formulating advice and counsel on system issues, assembly members frequently serve on working groups, policy committees, a personnel committee of general administration, as well as nominating members of the general faculty to serve on system initiatives as required.

The term curriculum refers to the means and materials with which students interact for the purpose of achieving identified educational outcomes. The curriculum that Etta works on is for college students. She also works on teacher preparation curriculums, serving on the teacher education council, which works in conjunction with Department of Public Instruction dealing with pre-K–12 curriculums. The State of North Carolina has developed the new core standard curriculum with science at every level. And teacher education curriculums are being revised to sync with the core curriculums. Teachers in North Carolina have to have a comprehensive science license for at least one subject. All K–12 teachers must teach some science. Every college student in the teacher education program at North Carolina A&T must enroll in "Methods of Teaching," whatever their specialties might be, prior to student teaching. Etta was a supervising teacher, from North Carolina A&T, for a student teaching in the chemistry department at Grimsley High School in Greensboro. This involved semi-monthly visits to observe, evaluate, make notes on, and coordinate with the cooperating teacher. She once was supervising a student who had already been offered two jobs.

Etta teaches many workshops for teachers with strategies and tips for motivating students and improving success rates. On the North Carolina A&T campus, the Center for Teaching and Learning conducts a variety of workshops designed to increase the graduation rate for university students. Currently, few specialize and remain in science. The center organizes activities to encourage students to remain in science, including workshops and professional conferences to share ideas and resources that will help center staff to help the students who come to them, especially those students who are underprepared, because the center takes underprepared students. Underprepared students are students who are in college but not yet ready to take the rigorous course of a college student for many reasons.

The university also has a Center for Academic Excellence that helps those students who have not decided what their major will be, or those students who don't have the prerequisites for the area in which they want to teach. One common scenario seen at this center is a student who comes to college, decides to major in, say, chemistry or chemical engineering, but never took the necessary math in high school. A weak math background or unwillingness to put forth the effort is frequently the biggest hindrance to student success in science. The Center for Academic Excellence a staff who can coach the students, motivate them, and develop strategies to help them succeed. Etta thinks that college teachers should have the same motivation and preparation that high school teachers have to get students interested in things that will help them be successful in their future careers.

In 1985–1986 when Etta was chair of the North Carolina A&T chemistry department seminar committee, she invited Dr. Ferguson to present a seminar on his research. He accepted, presented a seminar, and with his wife visited for several days.

Etta has been involved with the American Chemical Society ACS since her student days at Howard University, where she was a member of the Student Affiliates Chapter. After graduation, she became involved with Central North Carolina Section Chapter. She has served as chair of the hospitality group, treasurer, chairman-elect, and chair of the section. She has also been the past chair of the section and chair of the nominating committee. For the national body, she participated on the Women Chemists Committee WCC for ten years and the Undergraduate Task Force for programming at the national meeting.

As a member of the Undergraduate Programing Task Force she came up with ideas, themes, and workshops for students. These have included asking an eminent scientist to speak to the students, having the student chapters present workshops at national meetings, and getting other ACS members present workshops for students, especially career workshops. These activities require two to three years' advance planning. At the spring meetings, Etta ensures that student demonstrations are safe. She also has the various divisions meet with undergraduates at their poster sessions and present awards. She reviews undergraduate chapter reports, suggesting ratings for honorable mention, outstanding, or excellent—the student chapters receive their awards during the spring. She has served two terms on the task force and consulted with them for six years.

Etta was also involved with the National Organization for the Professional Advancement of Black Chemists and Chemical Engineers (NOBCChE) chapter in North Carolina, which started in the 1980s. She chartered a chapter and became president of the Greensboro chapter and served as president of the Piedmont Triad Chapter from 1992–1996 and vice president from 1997–1999. This chapter hosted the southeast regional meeting in Greensboro on November 12–19, 1992.

Etta has several publications related to classroom practices and strategies to her credit. She has initiated a technique called the inverted classroom. Students are given assignments to do outside of class. In class, they discuss their process and anything that was difficult. Thus, class time is more than just a lecture and becomes a time of sharing. Etta has taught chemistry and chemistry lab for nursing students and an online physical science class. She developed an online class and worked on an online advanced chemistry course as part of the master's of art in teaching in the School of Education. Some students take the online class by enrolling in North Carolina A&T and do their lab work in a community college near them.

Etta managed her family and career with a lot of support. While her career was evolving, so was her family. Etta and her husband had four children: Angela born in 1961, Clint Jr. born in 1965, Tessa born in 1969, and Latasha born in 1975. Etta spaced her children four years apart so that when one was coming out of college another would be going in. She has three daughters now, a son-in-law, and two granddaughters. Her son died of leukemia in 1976 when he was 11 years old.

As a member of Shiloh Baptist Church, she had a lot of support from other church members, who would babysit when she went back to college.

When her youngest child was born, she had a housekeeper who cleaned and took care of the baby. One reason Etta decided to go back to school was because her housekeeper was thinking about retiring. Etta decided to enroll while her home support was still available. In the evening when she was in class, her church members or mother would help with the children.

Because she was among the few African American students at the University of North Carolina graduate school, she consulted with people who had also attended about supportive advisors and thesis committee members. Having experienced problems themselves, they were very willing to help. Taking the initiative, Etta took non-credit courses and audited courses to get back into the habit of writing and studying. Professors who taught those courses were able to act as references on her application for the program at the University of North Carolina Greensboro which she did not get into because the person who was to approve it was on sabbatical. With the delay in the approval of her application she was urged to go to Florida to get the degree. Family made that option unfeasible. She persevered with the help of God and the caring and nurturing people at Shiloh Baptist church.

Over the years, Etta was also active at her church, chairing the Deaconess Board for about twenty-seven years. She also participates on the hostess committee and the administrative team, which advises the pastor, monitors the budget, and ensures bills are paid. She participated in the search for a new pastor and is involved with the Roy Lee Phillips Fellowship Club, which works to support the bereaved and hosts wakes in the fellowship hall.

Etta is involved with the State Employees Combined Campaign, an annual consolidation of funding requests from various agencies. State employees can designate how much money they want deducted from their pay each month to support these agencies or nonprofits. She is particularly proud that the year she chaired the effort saw an increase in employee contributions.

Etta is a fifty-year member of AKA. One of its projects, which she participated in, was Sister Walk of Breast Cancer, because one of her friends had the disease. Proceeds from the event funded a support-group chapter in Greensboro.

She is also a member and active volunteer of Links Incorporated, a group of African American women striving to increase the standard of living for African Americans and their children, looking at links to education, health and wellness, and so on. Etta was also a member of Jack and Jill, a mother's group that plans educational and cultural activities for children between the ages of two and nineteen. Active members have children still in high school or under nineteen. Etta's children have developed lifelong friendships from their participation in the group.

When asked about mentors, Etta mentioned several people from her time in graduate school at the UNCG: Dr. Dwight Clark, her advisor; Dr. Lois Edinger; Dr. Walter Sullivan, former chairman of chemistry at North Carolina A&T. She also credits the person who took her teaching job at Dudley High School when she had the opportunity to take up the grant at North Carolina A&T.

Etta considers her most important contribution to be the students she has taught who have become enlightened. They know the material. They feel good about what they have been taught. Numerous former students have returned to thank her for being so hard on them. They found they were well prepared and rose to the top of their class in other schools. Etta feels that is the greatest compliment that she could receive.

Etta loves teaching. She loves seeing that light bulb go off in students' heads when they say, "Oh I got it. I got it. I see what you're talking about." She especially likes teaching chemistry for nurses, helping them pass the state board exam.

She is also very pleased about being president of the honor society of Phi Kappa Phi, A&T chapter 291. This is a national honor society, over one hundred years old, with the motto, "Let the love of learning live within your heart. Learn to love." Begun in Maine, of its one hundred and forty chapters there are about four Black Colleges and Universities HBCU chapters. Both Hillary Clinton and Barak Obama are members. A number of Fortune 500 CEO's are also members and give away a lot of money for scholarships, study abroad programs, and literacy. Members are in the top seven percent of their second semester junior class; seniors or graduate student must be in the upper ten percent of their class. Faculty members must make some outstanding contribution to their field and be nominated by a member of the organization.

When asked about what she thought are the differences between the Jim Crow south she grew up in and the south of today, she said students now have more opportunities for advancement and growth. They also have more funding to continue their education. Students in high school now have clubs they can join. Also, teenagers can no longer drive school

buses. She thinks today's college students have no sense of time management because of bad habits in high school.

Etta's thoughts for future scientists? Do not let fear rule your life.
Think about being successful. If you think you can do it, you can. Also,
lead a balanced life of work, recreation, and socialization.[3]

3.2 Dr. Sondra Barber Akins

FIGURE 3.2. Sondra Barber Akins

Learning and Teaching Science: A Chemist's Journey

This story is in the first person because this was written by Dr. Akins

I am a science educator with a chemistry background and I am married
to a university professor who is a physical chemist. I was a teacher of
science and mathematics and an administrator in a New Jersey public
school district where I was responsible for implementing change in
curriculum and teaching practices. After I retired from the school district, I worked as an education professor at a New Jersey university
where I taught and supervised pre-service science and mathematics
teachers. Before I moved to New Jersey, I taught chemistry and physical
science in schools and colleges in Florida, Virginia, and Massachusetts.
I also worked as an industrial hygienist for an electronics company in
Massachusetts.

I was born March 16, 1944, in Winston Salem, NC, the third child of
Alexander Eugene Barber Sr. and Mabel Savannah Sharpe Barber. My
mother was born May 31, 1918, in Greensboro, NC. She was the second

child of George Sharpe and Lamar Ingram Sharpe. George and Lamar Sharpe had separated by the time Mabel and her brother Walter were ten and eleven years old. Lamar Sharpe moved to New York and she and George divorced. However, she returned periodically, and she was a strong and constant force in the lives of her children and her grandchildren. George's parents, Charlie and Mary Holmes Sharpe, helped to care for Mabel and her brother. Since he was working as a chef at a college in High Point, George decided it would be best to send Mabel and her brother to Mary Potter School, a reputable boarding school in Oxford, NC, where they could be together and get a good education. One of George's sisters was a teacher at Mary Potter. She looked out for Mabel and Walter and she kept George informed about his children's progress. Mabel Sharpe received a good education at Mary Potter. After Mabel graduated, her father sent her to Winston-Salem Teachers College (now Winston-Salem State University). He wanted her to be a teacher like three of his sisters. (My mother often told me that she never wanted to be a teacher. She wanted to marry a man, whom she loved and have children in a stable family where there was no divorce.)

My father, Alexander Eugene Barber Sr., was born on August 26, 1911 (or 1912) near Charlotte, NC. He was the second from the youngest of eleven children of John Barber and Mary Ivey Barber. After Alexander was born, the family moved to Winston-Salem to work in the tobacco industry. Mary Barber was widowed when Alexander was five years old and the youngest child was an infant. At the age of seventeen, Alexander was still living at home with his mother and attending school. He found a job very close to his mother's home at Winston-Salem Teachers College where he was allowed to live on campus, and he dropped out of the eleventh grade at Columbian Heights School. Alexander had an assortment of jobs on the college campus. In time, he became chauffeur for the college president, Francis "Frank" Atkins. (Francis Atkins was the son of Simon Green Atkins, the founder of the college, which began as Slater Industrial Academy in 1892.)

Winston-Salem Teachers College is where my parents, Alexander Eugene Barber Sr. and Mabel Savannah Sharpe, met in 1936. She was an eighteen-year-old freshman. He was in his mid-twenties and had been working on campus nearly eight years. Mabel saw Alexander eating peanuts and asked him to share them. He wouldn't. He told her he would bring her a whole bag of peanuts if she would meet him at the same location later. Their first date was on Valentine's Day 1936, and they married later that year. Alexander Barber became supervisor of the college laundry, and he opened a drycleaner business close to campus. In 1950, Alexander and Mabel Barber had completed their family of four children who were between the ages of one year and thirteen. (My oldest sibling is Alexander Eugene Barber; next is Freddie Walter Barber; Mary Barber Worthy is my youngest sibling.) That year, my mother began working in the laundry with my father. They were employed on the college campus during their entire working lives. He retired with forty years of service, and she retired with thirty years of service. When Alexander Barber Sr. died, at the age of 91, in 2002, my parents had been married for sixty-six years. Mabel Barber died in 2016 at the age of 98.

As a very young child, I occasionally visited the Winston-Salem Teachers College campus. But my visits became routine when I was six years old and my mother began working in the college laundry with my father. Since my brother Freddie and I were attending the afternoon session at Skyland School, my parents took us to the laundry with them. Each day before noon, we would leave the laundry, ride to our house, eat lunch, and then ride to Skyland School for the afternoon session. Then my parents would go back to work in the laundry. Shortly after Freddie and I walked home from school, my mother would be back home, preparing dinner. She didn't work at all during the summer when we had school vacation.

Being close in age, Freddie and I were natural learning partners. We had great fun together, at home, in the laundry, and in the home of my paternal grandmother, Mary Barber. My mother was delighted that we were good at math as well as reading. She would say her middle children took math ability "after" their father and they took it "from" their mother. Freddie was a year ahead of me in school. I remember trying to do the things Freddie could do. My mother said we competed with each other. As we got older, Freddie and I became familiar with other campus sites: dormitories, classrooms, the dining hall, post office, and the canteen. I talked to students who were learning to be teachers and nurses and I talked to instructors. Having that college connection was beneficial. I went to cultural events, homecoming parades, football games and basketball games, and I saw nationally known black celebrities who visited the college.

Our family home was in East Winston. In the beginning, there were only three houses on our street and there were fields and woods where we could play safely. My parents bought the house before I was born. In fact,

I was born in the house. Originally, we had only two bedrooms. Later, two bedrooms and a garage were added.

When I was little, my parents would take us to town. As we got close to town there was the sweet aroma of tobacco coming from the factories. I would marvel at the Reynolds Building that was erected by R. J. Reynolds Tobacco Company. (R. J. Reynolds and Hanes were the big industries in the city.) I always wanted my father to drive further so I could see the historic coffee pot that separated the old cities of Winston and Salem before they merged to become Winston-Salem in 1913. When I was older, I went to town alone, by public transportation. I never rode in the back of a bus because the black-owned Safe Bus Company took us from our segregated neighborhood to the black business area of town. Beyond the black business section of town is where I saw the words "colored" and "white" over water fountains and on restroom doors. I stood at lunch counters in Woolworths while whites sat down to eat in a different location.

Within our segregated, black community there was dignity and pride. There were successful people and thriving black-owned businesses. We had lawyers, and physicians who served black residents. There were college instructors and school teachers who lived in the community. There were funeral homes, taxi companies, hair salons, barbershops, a brick-making business, and a commercial school. We had a hospital with black doctors and nurses. A school for nurses was affiliated with the hospital. We had a thriving, black-owned insurance company. My favorite hangout was the East Winston branch of the public library where the librarians were black women who lived in my neighborhood. There was a black-owned radio station and a fire station with a black fire chief and black firemen. There were many black churches. My siblings and I grew up in Zion Memorial Baptist Church. My parents were among the charter members who founded the church as Second Mount Zion Baptist Church in 1944, when I was seven months old.

Economically, the black population in Winston-Salem, which was 43.8 percent of the total population in 1950, fared well, compared to most places in the south. Within the black community, there was an opportunity for professional and nonprofessional employment as just described. Black people were employed as laborers in white-owned industries, the most significant of which was R. J. Reynolds. As a result, black nonprofessionals owned and maintained nice homes and automobiles and they provided their children with all the essentials and more, especially when both parents in the household were working. There were some black people who were quite affluent and lived in impressive homes. Because we were all confined to living in segregated, black communities, my contemporaries could see what we could achieve, if we set our goals high and worked hard.

My school experiences were wonderful, in spite of segregation. Black teachers, principals, and other school workers were role models. Some of my teachers graduated from Winston-Salem Teachers College, so my parents knew them well. Many acquired graduate degrees from historically black colleges and others received graduate degrees from integrated colleges and universities in northern cities. We had books for all our basic subjects; a few were new; many were old. It was unfortunate that our books did not have pictures or stories of black people. That oversight was addressed every year during the second week of February when our elementary teachers and librarians taught about black achievers in education, literature, the arts, entertainment, and sports. I learned about George Washington Carver when I was in first grade. I always knew that black people had achieved and I could be an achiever too. I learned it from home, school, and church. All those institutions sent the same message to me. My favorite school subjects were language arts and arithmetic, when I was young. Science was not listed on my report card until I was in the seventh and eighth grades. However, nature was prominent in elementary activities, in the songs and poems, in the art activities, and in the fabulous plays and musicals at Skyland School. Leading to the school was a large park that was a nature trail unto itself. I remember looking at plants, collecting leaves, observing animals, and looking for shadows in the park and in my yard at home. My idea to do that must have started at school. I had exposure to science on the weekends when Mr. Wizard came on television. I envied the white children on the show. Like watching the other TV shows of the 1950s, it seemed like I was looking in on a white world. It bothered me.

I remember all my elementary teachers at Skyland School. My first-grade teacher was Miss Louise Smith. She graduated from Winston-Salem Teachers College and completed graduate work at New York University. My sixth-grade teacher was Mrs. Grace Hall. She also graduated from Winston-Salem Teachers College and she received her master's degree at North Carolina College (now North Carolina Central University). Mrs. Hall was very much into dramatics and history. Sixth grade is when students were placed in academic tracks. From sixth through eighth grade I was in the highest academic track. Miss Myra Rosemond, my seventh-grade teacher, was a taskmaster. She stressed problem-solving and told us it was time to take initiative and accept responsibility. She told us that one day we were going to be fine men and women, and we were going to have to live and work in an integrated world. That was the year when the Little Rock Nine integrated Central High in Arkansas, in the midst of hateful crowds. I didn't want to pay the price they paid to go to school with white children. I couldn't see how anyone could do their best work under those circumstances. But I knew I would have to learn with white children one day. I had been tuned into the civil rights movement since I was ten years old. That is when I heard about the Brown vs. Board of Education decision. Our pastor at Zion Memorial Baptist Church kept us informed about civil rights, and faith permeated practically everything we did that presented a big challenge or risk. Together, church, home, and school sent me the same message, that I could achieve and I should strive to become the best that I could be.

My eighth-grade teacher was Mr. Peter Roach, a Winston Salem Teachers College graduate who was from Boston, Massachusetts. One day, Mr. Roach did a physical science demonstration. He had a glass filled to the brim with water and an ice cube was floating at the top. He asked the class what would happen when the ice melted. I said the water would spill over. The ice melted and there was no spillover. I was embarrassed to be wrong because I was considered to be the top student. Knowing that science would be a serious high school subject, I became concerned, then and there. I had eight wonderful years at Skyland. My parents were involved in my education from first grade through high school. My father was president of the Skyland School PTA for two years. I graduated from Skyland as valedictorian in 1958. During the summer after I graduated, I wrote an editorial that was published in the Winston-Salem Journal. I responded to opinions black and white citizens were expressing about racial integration. I was insulted that some whites thought black people were intellectually inferior. I wanted to let people know that I went to school with smart children and our teachers were teaching us well.

My love of travel began when I was a child going places with my family. We went on picnics to the Blue Ridge Mountains. We visited relatives in North Carolina cities. My paternal grandfather lived in High Point where he was a chef at High Point College. We enjoyed going to see him. Once, we went to the college dining hall to see him, even though it was a college for whites. One of my grandfather's sisters lived in Greensboro where she taught school. Two of his sisters lived in Durham, which was a longer trip. One was married to a medical doctor and the other was a high school teacher. We drove to the country to visit the farm where my maternal grandmother's brother lived with his large family. During those visits, relatives often talked about us children, how well we were doing in school. During the summer when I was ten years old, we drove to the beach in South Carolina where I first heard about hurricanes. On our longest car trips, we visited family members in New Jersey, New York, Connecticut, Massachusetts, and Detroit. When I was a teen, my father and his friends started a travel club. The first trip I took with the travel club was to the Great Smoky Mountain National Park and Cherokee, NC.

I enrolled in Atkins High School in 1958. Atkins High was named for Dr. Simon Green Atkins, the founder of Winston-Salem Teachers College. Students outside the Winston-Salem city limits attended Carver School, which served students in all grades, instead of Atkins. (Two other high schools for black students, Anderson and Paisley, opened in 1960. While Carver was named for George Washington Carver, the other public high schools for black students, including Atkins, were named for prominent black Winston-Salem educators.) At Atkins High, my sphere of acquaintances expanded, because Atkins received students from four public elementary schools for black children (Skyland, Fourteenth Street, Kimberly Park, and Columbian Heights). A few Atkins students came from St. Benedict's Elementary School, a feeder school to St. Anne's Academy, the city's Catholic high school for black females. At Atkins, competition was fierce in the academically tracked classes. My classmates competed with each other. Competition was considered to be healthy. Most of us were close friends, inside and outside of school. We inspired each other to achieve.

I participated in many extracurricular activities at Atkins. I ran for student government (and lost). As a ninth grader, I was in the Latin Club and the Dramatics Club. I had a big role in the production *White Tablecloths*. I was in Junior Engineers from ninth grade through twelfth and I was elected president in my senior year. I was inducted into the National Honor Society as a junior. I was elected as president of our Honor Society during my senior year. I was impressed by our guest speaker who wore academic regalia and looked to be as young as we were. It registered with me that a person could have a doctorate without being a medical doctor. The Junior Engineers Club was my link to science. The club adviser was my ninth-grade science teacher. When I confided in him that science was my weakest subject, he invited me to the Junior Engineers meeting because he was the club advisor. I had a new friend who was in Junior Engineers and all my ninth-grade academic classes. He knew everything, including science, and he was only twelve years old. His father was a radiologist who brought new ideas to the school and community when he became PTA president. My awareness of possibilities beyond what I had yet experienced began to blossom.

My ninth-grade science project was sweet and simple, a demonstration of the crystalline structure of sodium chloride. I used yellow gum drops to represent sodium ions and green ones to represent chloride ions. The science project whetted my appetite to learn more. I wanted to know how scientists knew the structure of sodium chloride. I read something about x-ray crystallography, but I could not really understand it. I had not even taken chemistry, although I knew who the chemistry teacher was: Mrs. Inez Scales walked the corridors wearing a white lab coat. I was impressed when she gave a demonstration during which she "turned water to wine." I was impressed and eager to take her class. But I had to wait until eleventh grade

All my high school teachers took an interest in me, starting in ninth grade. My science teacher, Mr. Fred Parker, got me involved with the Junior Engineers. My ninth-grade algebra teacher was Togo West, Sr. (His son Togo Jr., who was two grades ahead of me, became the first African American to be appointed Secretary of the Army.) In Mr. West's algebra class, I sat in the front row, between two male students. One was a cool tennis player who became a state champion. The other was my brilliant twelve-year-old friend. Mrs. Clara Gaines was my Latin teacher. My ninth-grade English teacher and homeroom teacher was Mrs. Flonnie Anderson, the dramatics coach. I was most at ease with her during my freshman year. I tried out for, and landed, a leading role in the thespians' production of *White Tablecloths*. When I was in the eleventh grade, I finally had Mrs. Inez Scales as my chemistry teacher. It wasn't long before I knew for certain that chemistry was it. When I was in the eleventh grade, I took journalism from Mrs. Lois Woodland, and that is how I ended up on the publications staff and editor of the newspaper and yearbook.

During the summer following eleventh grade, I went to a science institute at North Carolina College (now North Carolina Central University) in Durham, North Carolina. At the summer institute I took mathematics and chemistry. The chemistry professor was Dr. Furth. He had a warm personality and a heavy German accent. I tried to learn everything he was teaching. The following year, when I was in twelfth grade, I took advanced chemistry at Reynolds High School, a school attended by white students. My advanced chemistry class whittled down to about five white students and me, the only black student. The class met several days a week. The students were cordial, nothing like what stayed in my mind about the integration at Central High in Arkansas. The class was taught by two men, Dr. Hounshell and Mr. Jarrell. My lab partner was a nice, quiet girl from Reynolds High School. I was outside my comfort zone because it was my first racially integrated experience. The subject matter pertaining to aqueous solutions and equilibrium was entirely new to me. But quitting was not an option. At the end of the year when our class went on a field trip to the state university, there were so many white college students surrounding us that I took comfort in being in the small circle of advanced chemistry class students.

My performance in the advanced chemistry class was okay. I survived. My strongest support came from one of my Atkins classmates who was taking the advanced biology class that was also held at Reynolds High School. She was a close friend and the only black student in her advanced biology class. Our fathers took turns driving us over to Reynolds High. We gave each other moral support. Other support came from my physics teacher. After having advanced chemistry in the morning, I returned to Atkins for my afternoon classes. Mr. Campbell, my physics teacher would ask me how things were going. He was laid back and had a way of taking the edge off things. It seemed that he knew, or knew of, one of the teachers of the advanced chemistry class. In spite of the challenge of AP chemistry, I still wanted to major in chemistry in college. I graduated as valedictorian in 1962, and I was offered a full-expense scholarship to Howard University, which happened to be my first choice. (While I was in high school, guidance counselors were encouraging students to apply for admission to white colleges. I knew two students in the class ahead of me who went to state universities. One student in my class went to a smaller white college in Guilford County. He became a science teacher. There were other students in my class who majored in sciences.)

In 1962, I entered Howard University as a participant in the University Honors Program. It was a wonderful opportunity that set the stage for the rest of my life, although I completed only one academic year there. I lived on the honors floor of Baldwin Hall, where there were excellent role models. One was a bright young woman in pre-law who had been the valedictorian of the Atkins class of 1959. There were math and science majors among the students on our honors floor. It helped to have girls from my floor taking the same chemistry class I took because we could talk about the class in the dormitory. We got in a large chemistry lecture class taught by one of the fiercest instructors who was known for assigning many failing grades. At Howard, I was required to complete an honors project. I talked to a pre-med male student from Durham, NC, and asked him for suggestions. He immediately introduced me to his friend, Daniel Akins, a senior chemistry major. Daniel Akins asked if I was interested in x-ray crystallography. Of course, that struck a chord because of my ninth-grade science project. He walked me to the physics department and introduced me to the chair of the physics department. The physics chair introduced me to Dr. Sohan Singh, who mentored me through the project for a full semester. Working with Dr. Singh, I had exposure to upper class and graduate students who were extremely focused. They practically lived in the laboratory. I was comfortable in that environment. My x-ray crystallography project confirmed that chemistry was the right major for me. At Howard, there were many black women who were going into medicine. But I knew, back in high school, that I didn't want to work on sick people who might die.

As it turned out, Daniel Akins and I married on August 21, 1963. He had already been accepted into a PhD chemistry program at the University of California at Berkeley. In September, he went to Berkeley and I went back to Howard and started my sophomore year taking organic chemistry from Dr. Lloyd Ferguson, who had gotten his doctorate in chemistry from Berkeley. Within a month, my husband and I decided, together, that it would be best if I joined him at Berkeley instead of finishing the academic year at Howard. The dean of women at Howard University tried to talk me out of dropping out of college. My parents did not even try to talk me out of leaving Howard University. From their own experience, they understood. (I had learned their story directly from them.) They believed husbands and wives should be together, and they were supportive. But they wanted assurance that I would finish college. In fact, during my first year at Berkeley, when I had to pay out-of-state tuition, my parents paid it. That is how understanding and supportive they were, even though I had given up a four-year scholarship when I left Howard. I went home to Winston-Salem to spend several weeks with my parents before I boarded a Greyhound bus to California. To this day, I am grateful for my year at Howard. Living on the campus, I became familiar with African American history and I made friends with whom I reconnected in later years. I am grateful that my parents had confidence in me and never displayed any disappointment when I left Howard.

Among the first people I met at Berkeley were my husband's research advisor and his wife. I was surprised by the fact that Professor C. Bradley Moore, who had finished his PhD under Professor George Pimentel, had gotten his doctorate the previous year and was only two or three years older than my husband. That year, my husband was Professor Moore's first, and only, graduate student, although the research group grew, considerably, during the following years. Professor Moore's wife, Penny, was an undergraduate majoring in physics. Often, she worked long hours in the laboratory, along with her husband. Since my husband worked long hours in the laboratory, I spent a lot of time there too. From the beginning, there was a friendly and collegial relationship among us. We were all on a first name basis. We have remained friends to the present day.

I enrolled in the College of Chemistry at UC Berkeley in January, 1964. The first time I walked through the campus, I was amazed by the political activity of the students. There were tables filled with paper and students were getting people to enlist in various activities, including civil rights in the south. I felt uneasy that I was not involved in the activities. But I was reminded that, by continuing my education in that competitive environment, I was contributing to the civil rights struggle. There was no time for any undergraduate or graduate science student to do anything except go to demanding lectures, take laboratory courses, and study for formidable examinations. My husband and I had two goals: A degree (PhD) for him and a degree (BS) for me. Not one without the other. My program was in the College of Chemistry where I was getting a bachelor of science, rather than a bachelor of arts. That meant a heavier concentration of chemistry, physics, and mathematics courses and fewer liberal arts courses. Since I had satisfied most of the liberal arts requirements at Howard, I spent most of my time in the science buildings.

While I was at Berkeley, I knew only a few black students. I remember only one other female student whom I could identify as black in one of my huge undergraduate chemistry classes and one black male student in my undergraduate physics class. Sadly, we never talked and I didn't know their names. If it had been years later, when the numbers of black students increased, there would have been black student organizations and there would have been more camaraderie. I became friends with a black woman who was majoring in biology. She was married and had a young child and changed her major and transferred to San Francisco State. I don't really know whether a black woman got a degree in the College of Chemistry before I arrived. I am certain there were no women on the chemistry faculty. I remember meeting six black males, besides my husband, who were graduate students in chemistry. The wife of one of those students had a chemistry degree from Spelman College, and she worked at the Lawrence Radiation Laboratory. During our last year at Berkeley, a black couple arrived to work on their doctorates. He was a Morehouse College graduate, and she was a Spelman graduate.

On March 3, 1965, my husband and I had our first child. I had two years in undergraduate school still ahead of me. I took the spring semester off to have the baby, and I got back into school during the summer sessions when I completed physical chemistry. The following year, I worked in the storeroom twelve hours per week and paid everything I made to the baby-sitter who lived across the court from us in student housing at University Village. We lived on the monthly stipend of two hundred dollars which my husband received as a research fellow. Berkeley was a tough university for an undergraduate in science. In the chemistry and physics classes I took, the professors graded on the curve and the class average on tests was a C. In spite of that, I never even considered changing my major. I did my senior research project under the supervision of Professor C. Bradley Moore, in the same laboratory where my husband was working. (Dr. Moore and other graduate students in Dr. Pimentel's group had been among the first Berkeley scientists to work with lasers.) I was in the midst of excitement and discussion about what had gone on in earlier days at Berkeley. I felt comfortable with respect to the theory underlying molecular spectroscopy by my senior year. I received my bachelor's degree in chemistry in June 1967, and I continued working in Professor Moore's laboratory through the following academic year as a paid laboratory technician. My job paid more than twice my husband's stipend. In June, 1968, my husband received his PhD in chemistry. Within five years, we had achieved what we promised ourselves and our parents. I had gone back to the east coast to visit my parents only once. My husband had not seen his parents at all during that time. We flew back to the east coast right away with the baby girl they were eager to see.

In September 1968, I began a program for a master's degree in chemistry at Florida State University in Tallahassee, where my husband had a post-doctoral fellowship with Dr. Michael Kasha. (Dr. Kasha had been one of the last graduate students of G. N. Lewis at Berkeley.) I had always favored physical chemistry so I took graduate courses in quantum mechanics, thermodynamics, and statistical physics. To broaden my background, I took inorganic chemistry courses and worked in the laboratory of Professor James Quagliano where interesting solution chemistry research was being conducted. That is how I met Dr. Lidia Vallerino, a scientist in her own right, who was the wife of Dr. Quagliano. She had co-authored the book *Chemistry: A Humanistic Approach* with her husband. However, she did not have a regular faculty appointment. She was the mother of two boys, a scientist, and the wife of a scientist. I admired her accomplishments as a scientist, mother, and wife of a scientist. We became friends. I felt I had made the right decision to pursue a job with emphasis on teaching rather than research. Part of that was due to the fact that all the laboratory work I had to do at Berkeley after our baby was born had taken its toll. I didn't want to work long hours in the lab and be away from my child, and I really favored teaching. As a teacher with insights about research I would be even more effective.

After I received my master's degree from Florida State, we moved to Tampa, FL, where my husband became an assistant professor at the University of South Florida. That year, I took a job at Greco Junior High School teaching introductory physical science. Early on, I had to confront the reality that thirteen- and fourteen-year-old students had lots of things besides learning science on their minds, and I also realized that I did not have the temperament or preparation to teach these children and adolescents. During the following year, I was a chemistry instructor at St. Petersburg Junior College in Clearwater, FL. It was a perfect match for me. Three years later, I was expecting our second child and I wanted to work closer to home. I was fortunate to get a similar position at Hillsborough Community College in Tampa. Three years later, my husband accepted a two-year job at the National Science Foundation (NSF). I became a chemistry instructor at Northern Virginia Community College in Annandale, which was close to our home. I enjoyed teaching in two-year colleges because my students were mature adults. Some were planning to transfer to chemistry programs in four-year colleges, so I also had the opportunity to teach conventional first-year chemistry courses. I learned about fire science, forensic science, and health professions in order to teach according to my student's needs and interests. I became interested in integrating science with the social sciences, the arts, and history. I took different courses and went to cultural events that complemented the rigorous science background I had acquired in undergraduate and graduate school. Those were, indeed, my golden years of teaching. I was learning and my learning was stress-free. I was able to manage working outside the home, because I had built stamina back at Berkeley. I even had the means to pay for help with housework, if I needed it.

When my husband's two-year position ended, he decided to work as a senior scientist at Polaroid in Massachusetts. I was open to the adventure because I believed it would be easy to find a job with a master's degree in chemistry and nine years of teaching experience. Upon arriving in Massachusetts, I got a job, immediately, as a replacement for a physics teacher on one-year maternity leave from the high school in Lexington. My students were very serious and many were bound for Ivy League colleges. After that, I enrolled in a doctoral program in chemistry at Brandeis University. Then, the position of industrial hygienist, something different from anything I had ever done, was advertised by the Hewlett Packard Company in Waltham. I took a leave of absence from the doctoral program and took the job at Hewlett Packard. I was a novelty there. Workers and management had not seen anyone like me. The culture was completely different from academia. While working, I took courses from Harvard School of Public Health and Liberty Mutual Insurance Company. I worked under the direction of the corporate industrial hygienists in Palo Alto, CA. Within two years, I was instrumental in implementing an industrial hygiene program at the Waltham division that received a passing grade from the corporate office monitoring visit. While at Hewlett Packard, I updated the Toluene Hygienic Guide that was published by the American Industrial Hygiene Association in 1984. I had a great feeling of accomplishment and, seemingly, a bright future at Hewlett Packard, a company that was considered the best place to work. It was not to be, because a family relocation was necessary. My husband had made a sensible switch from his job at Polaroid to a position of full professor of chemistry (with tenure) at City College in New York. It had required him to commute between New York and Massachusetts for two years. That was not the lifestyle we wanted, so when our house in Massachusetts sold after two years on the market, I gave up my job at Hewlett Packard and moved. I was unhappy about giving up the job. I didn't know what the next adventure would bring.

In 1983, we moved to Englewood, NJ, a city I fell in love with almost immediately. I felt good about the racial diversity of the city and its schools, something that was missing in all the places we had settled previously. Dwight Morrow High School was a five-minute walk from the house we bought. It was about to open in September without coverage of some of the math classes. I took a job as a mathematics teacher, thinking it would be for one year. I stayed another year to teach physics. The next year, chemistry courses, included the new offering of advanced placement chemistry, were in need of coverage. Chemistry was my first love for learning and teaching. So, I stayed. After teaching for five years in Englewood, I had acquired permanent certification for teaching science and mathematics and for administrative positions.

In 1988, I became the district supervisor of mathematics, science, and technology. There were critical issues in education we needed to address. In comparison with other advanced countries, students in the United States were underperforming in science and mathematics; women and minorities were underrepresented in science and mathematics courses and careers; there was a dire need for teachers of science and mathematics and most elementary teachers were lacking in sufficient content knowledge and pedagogy to teach science and mathematics effectively. In addition to those problems, feedback from the workplace indicated that college graduates did not have the kind of skills that would be needed by workers during the twenty-first century. All of this indicated that we needed to make changes in the way we were preparing students. My job as an administrator was to facilitate change in science and mathematics curriculums and teaching practices across the district. I found it stimulating to work toward solving those problems. I knew my prior experience would serve me well in that effort. In order to bring about change we had to have broad partnerships including parents, higher education schools, and businesses in the community. We also needed funding. I became a liaison between the district and various partners. Beginning in 1990, we had a three-year collaboration with Ramapo College of New Jersey, funded by the US Department of Education, and a partnership with Corn Products Corporation (CPC) in Englewood Cliffs. The backbone of the Ramapo College partnership was ongoing professional development for elementary teachers in science and mathematics, which enabled them to improve their content knowledge and pedagogy. Ramapo College mathematics and science professors helped to identify science and mathematics concepts that should be emphasized throughout the K–12 curriculum. We worked on curriculums, a plan of what should be taught across the grades, in those subject areas. Emphasis was on building concepts within the disciplines and integrating concepts across the disciplines. Continuous workshops for teachers ran through the year and summer months. We brought the elementary and high school teachers together to develop exciting outcomes for high school graduates that would be supported throughout the K–12 curriculum. We put other resources in place, such as science and mathematics subject-area leaders and offerings of family science and family math throughout the elementary schools. Science and mathematics professors from Ramapo College made regular visits to elementary classrooms and worked with the teachers and children, so the children got to know scientists and mathematicians personally. Elementary school children went to Ramapo College where they saw the scientists and mathematicians at work. Through our CPC partnership, elementary teachers went to workshops in industry.

Everyone was learning and applying new learning to teaching. I was no exception. When I became an administrator in 1988, I also enrolled in a doctoral program at Teachers College of Columbia University. The degree for this program was an EdD (Doctor of Education) instead of a PhD I would have gotten as the terminal degree in a graduate chemistry program. I tailored my doctoral program so that every course I took, and the research I did, was totally relevant to the work I was doing as an administrator. My dissertation, "Restructuring the Mathematics and Science Curriculum: Elementary Leadership Teachers," was completed in 1993. I targeted elementary teachers because that is where it should all start. If students have early exposure to science and mathematics concepts, and if subject matter is taught effectively throughout the grades, then students' understanding will deepen as they move up the grades. If they don't have essential prior knowledge, there is little to build on.

After working as an administrator for eight years, I was jolted when my administrative job was abolished due to fiscal considerations. I returned to teaching in the high school, using new ideas and strategies that I had promoted during my years as an administrator. For the next two years, I taught integrated science at Dwight Morrow High School. I also taught chemistry in the Upward Bound program at Ramapo College. Then, the principal of Dwight Morrow High School left just before the 1997–1998 school year began, and I became principal of Dwight Morrow High School for fifteen months. After that, I served as staff developer and taught mathematics until my retirement from the school district in 2001.

I had the opportunity to participate in statewide efforts to improve curriculum and the quality of teaching while I worked in the Englewood school district: The Governor's Teaching Scholars Committee sought to revitalize teaching in New Jersey by awarding grants for college tuition to high school graduates in the upper fifth of their class. The NJ Science

Curriculum Frameworks Committee identified the science content and skills to be emphasized from first grade through high school in state schools.

I made presentations at the Science Convention and served on the program convention committee. From 1996 through 2006, I taught science and mathematics teachers in the Northern New Jersey Provisional Teachers Consortium. Through the consortium, persons who had previous science and mathematics careers in industry and other venues could teach in schools while earning permanent teacher certification. Beginning in 1993, I worked with the New Jersey Statewide Systemic Initiative (NJSSI). This effort was funded by the NSF with the five-part goal of enhancing education for all students; producing standards-based curriculums; fostering cooperation among schools, universities, and industry; gaining community support and aligning education policy; and increasing teacher command of subject matter.

I also served the Teaneck-Englewood community as mathematics and science consultant to the African American Educational Center of Northern New Jersey. This center was started by black parents of Teaneck, who supplemented their children's regular school education through a Saturday School which promoted general achievement and black cultural enrichment in the arts and sciences. The center was sustained for twenty-five years. When it closed in 1999, its entire curriculum was archived in the collections of the Schomburg Center for the Study of Black Culture.

In 2001, I began my fourteen-year tenure as a professor in the College of Education at William Paterson University. I taught science and mathematics methods courses and I supervised science and mathematics student teachers in urban and suburban schools. I participated in funded projects and I worked with faculty in the College of Education and the College of Science to secure and maintain the William Paterson University program for the certification of science teachers. This required adherence to rigorous requirements of the National Council for the Accreditation of Science Teachers (NCATE) and the National Science Teachers Association (NSTA). An important part of my work was recruiting candidates for the science teacher certification program. In order to acquire science teacher certification, all the candidates are required to complete a regular science major as well as a planned program of coursework in education and practicum and student teaching in schools. It remains a challenge to attract enough science majors into teaching when there are many alternative careers they may select in medicine and research. So, I worked with the

science faculty and I talked with students in the College of Science to ensure students were aware of teaching opportunities, as early as possible,
during their college studies. In time, we developed the program so undergraduate students have a double major and receive degrees in in their
science major and in education. During my last year at William Paterson,
we instituted a master's education program for candidates who already
had degrees in science. I planned the science methods course for those
students and taught the first cohort of those students.

As a science education professor, I combined scholarship and teaching.
I published a paper, "Bringing School Science to College: Modeling
Inquiry in the Elementary Science Methods Course" in a NSTA mono-
graph, *Exemplary Science: Best Practices in Professional Development*. This
2005 publication featured fourteen exemplary professional development
efforts from across the country which addressed the National Science
Education Standards adopted in 1996. My paper described my rationale
in developing and teaching the elementary science methods course, my
teaching practices, and the responses of my students who were preparing
to become elementary teachers. I also made presentations at teacher education conferences and science education conferences. One of my most
gratifying presentations was that of the keynote speaker at the Minority
Affairs Committee (CMA) luncheon during the 2008 national convention
of the American Chemical Society ACS: "Achieving Equity and Excellence
in Science: Looking Back and Moving Forward." I spoke to professional
chemists who were celebrating the fortieth anniversary of the society's
Project SEED program and hundreds of pre-college students who were
participants. Based on my learning and teaching experiences, I had the
opportunity to share my thoughts about how students can be helped to
achieve: Students should learn from teachers who believe their students
can achieve and inspire students to become the best that they can be.
Students need to receive positive, reinforcing messages from home,
school, and community. Students should have positive role models at
school and in the community. Students should work together; they should
support and inspire each other. Students should know they are central to
our vision for the future.

I share reflections about my learning and teaching because I want to ac-
knowledge the people and practices that made a difference in my life. Also,
I hope others will realize the opportunities in teaching science and the
rewards that can be derived from teaching, in general. Majoring in chemistry was the right choice for me. I am amazed that, back in high school,

I was introduced to a field that was basic to a forty-five-year journey of productive work. By being open to lifelong learning, I had several careers, all of which were science related. Choosing teaching over research was the right decision for me because it was most compatible with the lifestyle I wanted. The bonus was being able to integrate my work and personal life so they were mutually enriched. Our children and grandchildren had a heavy exposure to science. It was no surprise that our first daughter (Dana Akins-Adeyemi) became a mechanical engineer. She married an electrical engineer whom she met in graduate school. Their older daughter is a chemical engineer and their son is majoring in engineering. Their younger daughter just finished middle school. Our younger daughter (Meredith Ivy Akins) had a twenty-year career as a Broadway dancer. She married an actor and she is the mother of a two-year-old son. Now she is the theater teacher in a school for children with learning differences.

3.3 Saundra Yancy McGuire

FIGURE 3.3. Saundra Yancy McGuire

Dr. McGuire (Fig. 3.3) is the Director Emerita of the Center for Academic Success and retired Assistant Vice Chancellor and Professor of Chemistry at Louisiana State University (LSU). She developed a process to help students succeed in science and has mentored hundreds of minority students to successful careers. She is the author of *Teach Students How to Learn: Strategies You Can Incorporate into Any Course to Improve Student Metacognition, Study Skills, and Motivation* and *Teach Yourself How to Learn: Strategies You Can Use to Ace Any Course at Any Level*, and she now lectures around the nation about these strategies.

Saundra's mother, Delsie Melba Moore, was born in Maringouin, LA. After her family moved to Baton Rouge, she went to high school at Southern University Laboratory School and then went on to Southern University A&M College where she met her future husband, Robert Ernest Yancy Jr., while studying to become an elementary school teacher. She did not work, except occasionally as a substitute teacher, while Saundra was growing up.

Saundra's father was the oldest male in the family of nine children born to Robert Ernest Yancy Sr. and Effie Jane Gordon Yancy and grew up in rural Greensburg, LA. Saundra's great grandfather on her father's side, Isaac Gordon, was a freed slave. All of the children in Saundra's father's family went to college and all but two received graduate degrees. Saundra's paternal grandmother went to Leland College in Baker, LA—one of the oldest private black colleges in the country— and became a teacher. She decided early on that all of her children would go to college. Seven of the children have master's degrees from top national universities, a result of the state's reluctance to integrate LSU and instead paying for black students to pursue graduate degrees elsewhere.

Both of Saundra's parents were born in 1918 and were undergraduates at Southern University between 1937 and 1947. Her father's college studies were interrupted by his service in the army during World War II; he only completed his master's in 1957 at Iowa State University, because he taught during the academic year and could only continue his studies in the summer months. At the time black teachers, unlike their white coworkers, were not allowed to take classes at LSU. Committed to furthering his education with an advanced degree, and knowing this would provide more resources for his growing family, he persevered for seven years. Her father taught agriculture and general science at Caneyville High School in Zachary, LA.

Saundra's grandmother's strong education ethic and work ethic were transported through her children to the thirteen grandchildren in Saundra's generation. Saundra has an uncle with both a master's degree and an educational specialist degree from Michigan State University; an aunt with a master's degree in home economics from Kansas State University; and an aunt with a master's degree in education from the University of California (UC) at Los Angeles. Among the thirteen grandchildren, there are four MDs, two PhDs, five with master's degrees, and one with a bachelor's degree.

Saundra knows that the story of her father and his siblings is very unusual. First, it's very unusual for the daughter of a slave to have nine college-educated kids, seven of them holding master's degrees. Second, the way they did this is unusual. Her great grandfather, the former slave, acquired some land in Greensburg, about an hour from Baton Rouge. Her grandmother inherited that land after he passed away and was able to parley it into a loan from the bank in Greensburg. With the loan, she bought some property in Baton Rouge and built a house. All of her aunts and uncles who went to Southern University actually stayed in that house. At that time, many students who went to black colleges usually worked on school farms or on campus to earn their tuition. Her uncle—the educator who went to Michigan State—played football at Southern University and had financial aid. The students would send whatever money they could home to Greensburg to help Saundra's grandmother with the expenses for their younger siblings. And when they also went to college, they did the same. Yes, it was very, very unusual.

Saundra was born on December 13, 1949, second child and first girl in a family of four children. She and her siblings grew up in Baton Rouge and all attended Southern University, a Historically Black College or University (HBCU), whose main campus is in Baton Rouge. She is very proud of her siblings and their achievements. Her brother Robert Ernest Yancy III, a year and a half older than she is, is also a chemist by training. He has a master's degree from the University of Arizona and works as a Food Safety and Quality Assurance consultant/auditor. Her brother Eric Albert Yancy, three years younger than she is, double majored in chemistry and psychology and now is a pediatrician in Indianapolis, IN. He went to Creighton University Medical School and was named alumnus of the year in 2011. Saundra's younger sister, Annette Louise Yancy, majored in psychology, received a master's degree from Western Michigan University, and is now an academic advisor at LSU. She was named the National Academic Advising Association (NACADA) Advisor of the Year for 2012.

Saundra grew up in north Baton Rouge, LA, about a mile and a half from the campus of Southern University. Although she grew up in the Jim Crow south, her parents had a wonderful way of protecting their children from its problems. Saundra and her siblings knew they couldn't drink from certain fountains or go to certain stores. But their parents explained this as coming from the ignorance of the people instituting those policies and having nothing to do with anyone being inferior. Saundra grew up with a healthy sense of self-esteem, never thinking she was better than anybody else, always knowing she was just as good.

Saundra thinks that she had it a lot better as an African American child growing up than youngsters growing up in today's integrated society. Today's children often don't have teachers who are seriously invested in their success. Although things were very segregated at the time, she thinks the education was better for black students in many ways. Wonderful teachers, who were also very nurturing and supportive, prepared students for the challenges that they knew were ahead. Families and schools worked closely together to help their students. She had a wonderful time at Southern University Laboratory School, which she attended up to the eleventh grade.

Baton Rouge schools were integrated in 1963, when Saundra was in the ninth grade. Because of the resistance to integration, the district decided to integrate one grade per year starting with the twelfth grade. Therefore, the first year she was eligible to integrate was in 1965 when she was in the eleventh grade. She decided to join the relatively few black students who would attend the previously all-white Glen Oaks High school in her junior year. She did not have any trouble academically because she was very well prepared. There were only two black juniors and a total of eleven black students in the whole school, with a student population of eleven hundred. Only juniors could take chemistry, and the administration would not allow the two black juniors into chemistry. They were told there was no lab space. So, the two juniors were placed in physics. She guesses there was some hope they would flunk out. She did well in physics, and she started at Southern University, after skipping her senior year of high school, without having taken high school chemistry.

Going to Glen Oaks integrated school was torture. Baton Rouge did not want to integrate, and the white students created a very hostile environment for black students. The white students would write things on Saundra's locker like "Damn Commie nigger." They would put razor blades on the tips of their shoes and try to kick at black students. The year Saundra integrated was the third year of integration (1965–1966), but since the authorities were going to integrate one grade at a time it would take twelve years to integrate the whole system. They decided to start with the twelfth grade because they did not want it to work and believed that if they started with little kids, they would get along with each other. Integration was voluntary then. Community organizers, including the American Friends Society (Quakers) came and recruited black students to integrate. Saundra's brother had integrated the year before her, but he also skipped twelfth grade so he was not there when she attended Glen Oaks. She does not feel bad about skipping her senior year because she had no social life at all in the school. She was not allowed to participate in any of the extracurricular activities. Most of the black students who went to the integrated schools still maintained their social connection with the black schools that they had attended.

In 1966, Saundra started at Southern University as a sixteen-year-old freshman. Southern University was founded by Joseph Samuel Clark in 1880, and Saundra's church, Mt. Pilgrim Baptist Church, was founded in 1893. Southern University was the largest HBCU in the world at the time she was there. In the 1960s, there were three campuses with a combined population of over ten thousand students: Southern University Baton Rouge, Southern University New Orleans, and Southern University in Shreveport. Since then a Southern University Law Center has been added.

Saundra credits part of her success to the Southern University Laboratory School she attended. Some of the students had come from other all-black schools (e.g. Capitol High School and McKinley High School) with outstanding reputations and excellent academic programs. Many of the students had parents who were faculty members at Southern University or educators in the school system. Many are now in education, politics, or the health professions. In the 1950s, Baton Rouge had many thriving black businesses, three black theaters, and several black hotels. Only within the past forty years, after integration, has that the community deteriorated to the extent it is now. When Saundra went to school many of her teachers had master's degrees. Her husband, who grew up in New Orleans, LA, had several high school faculty members with PhDs, who had gone to schools like Columbia University in New York City or the University of Michigan, yet it was hard for them to get positions other than high school teacher.

How did Saundra get into Southern University without a high school diploma? The university had a chemistry institute, sponsored by the NSF, for students who had completed their junior year of high school. At the end of the six- to eight-week program, students were invited to take the university entrance exam. Any student scoring within the top ten percent of the scores of entering college students was invited to skip their senior year and enter college early. Saundra did not go to the chemistry institute, because she had not taken high school chemistry, but she did go to the mathematics institute at Florida A&M University. When she returned from that institute, she asked her father to ask the director of the chemistry institute, Dr. Vandon White, if she could take the test and he agreed. She was a happy camper when she learned she did not have to take her senior year at Glen Oaks.

Saundra was able to live on campus at Southern University even though her home was only a few minutes away. Her parents wanted her to have a real college experience. Even though she was only sixteen she was very tall, and most people did not realize she was that young. Her close friends and roommate knew; it was not an issue for them.

Her freshman year was one of the best years of her life. A participant in the honors program, she was invited to a party by the program director. She'll never forget the date—October 13, 1966. That's where she met her future husband, Stephen C. McGuire.

Her roommate, Patsy Lawrence, was also a chemistry major, so they studied together. She had a lot of fun, despite the rules that girls were not allowed to ride in cars or wear slacks, and the only entertainment was movies. Every Friday and Saturday night there were movies on campus. A young man would to go to the dormitory office, the person at the desk would call the young lady's name, and the young lady would go to office to meet her date for the night. Curfew was 10:00 PM every night. There was a weekly vespers service on Sunday evening at 7:00 PM, and everyone had to attend. Vespers was wonderful, because fantastic, nationally known speakers came to speak. Saundra was also involved in the Baptist Training Union, a Christian organization

Ironically, Saundra majored in chemistry even though she had never taken high school chemistry. Dr. Vandon White, the chair of the chemistry department at the time, had a great faculty of fourteen or fifteen black PhDs, including Dr. William Moore, Dr. Wilbur Clarke, Dr. Press Robinson, and Dr. Sydney A. McNairy Jr. There were also women faculty members: Dr. Mildred Smalley, Ms. Celestin Tillotson, and Mrs. Willa

Moore. Saundra majored in chemistry because, when Dr. White asked her what her major would be, she said math or science. He asked her to consider a major in chemistry and she agreed. She really began to love chemistry when she was able to spend time in special programs designed to help black students excel. Since she was in college from 1966–1970 there were not that many blacks with PhDs who could hold faculty positions in institutions. But there were programs designed to increase this number by recruiting and preparing students for those positions. The summer after her sophomore year she spent at Columbia University in New York City in the Harvard/Yale/Columbia Intensive Summer Studies Program designed for students the coordinators felt had the potential to be college faculty. They provided scholarships to students who could study at either Columbia, Harvard or Yale for the summer. Saundra and her future husband were both selected to be in the program and went to Columbia University.

In her junior year, environmental pollution was discovered in Alsen, LA, caused by Crown-Zellerbach Corporation's paper mills. Part of the settlement for cleaning up the waste included provision of scholarships for two students from Southern and two students from Grambling State University to study in California in their junior year. Crown Zellerbach's headquarters was in San Francisco, so the three campuses available were UC Berkeley, UC Los Angeles (UCLA), or Stanford University. Saundra and Steve were again selected as the female and male students from Southern. She decided to go to UC Berkeley because it was number one in chemistry at the time and Steve went to UCLA to study physics. After Berkeley, she participated in the Harvard/Yale/Columbia Program for an additional summer and went to Harvard. During her senior year at Southern she decided she would go to graduate school after graduating in 1970. She was awarded a Danforth Foundation Fellowship to attend graduate school, which allowed her to go anywhere she chose. She decided to go to Cornell University in Ithaca, NY. The Danforth Foundation, a philanthropic organization, wanted to reduce the number of HBCU departments not accredited due to insufficient faculty with terminal degrees. At the time, they gave fellowships for students who wanted to teach in college. Their initiative involved actively recruiting students at HBCU's with the potential to complete a PhD and teach at black colleges.

Saundra arrived at Cornell during one of the coldest winters ever and found she was required to be a TA during her first year. She taught two classes and fell in love with teaching. She decided to change her PhD study from chemical research to chemical education and switched during the second semester of her first year in graduate school. She obtained a master's degree in chemical education from Cornell after one year and a summer working with Dr. Joseph D. Novak—well known in the science education community because his team developed the idea of concept mapping. He introduced Saundra to the notion of teaching students *how* to learn. She was fascinated by how a teacher could build student confidence, especially in those students fearful of chemistry, by giving them strategies and explaining things in terms they understood. At Cornell, she taught freshman chemistry and had some wonderful students in pre-med or other pre-health professions. She received a teaching award at the end of her first year. She received her master's degree in 1971.

Also at Cornell, she began helping students develop better learning strategies for chemistry. As a graduate teaching assistant, she was required to sit in on the lectures of the courses for which she was a TA. It occurred to her that it would be hard to learn the information from the way the lecture was presented. Since Cornell had a lot of minority students who came from schools in New York's inner city New York, Saundra went to the Minority Student Programs office—called COSEP (Committee on Special Educational Projects)—to offer to hold review sessions for any student who wanted to attend, which she held at the Africana Studies and Research Center on Sunday evenings from 7:00 until 8:30 PM. She translated for the students what was being said in the lecture into language they knew and could understand. She also gave them "between the lines" information about the things they were expected to know but that were never described in lectures. She had about forty to fifty students who would attend on a regular basis, and they had a good time together. They were able to do well on the exams, and many of the students are successful today. At the time there was no formal learning center at Cornell.

Saundra married after leaving Cornell, and the couple lived in Rochester, NY, where Steve was a physics graduate student at the University of Rochester. After four years they moved back to Cornell, where her husband pursued his PhD. Saundra found an established university Learning Strategies Center at Cornell, conducting weekly review sessions in general chemistry, physics, calculus, biology, and economics. She was able to obtain a part-time position there and spent mornings with her daughter; she also taught some chemistry courses, including organic chemistry.

When her husband finished his PhD in nuclear science at Cornell he obtained a position at Oak Ridge National Laboratory as a physicist, and they moved to nearby Knoxville, TN. It was there she learned that the University of Tennessee at Knoxville had a very strong science education program. So, she entered it to work on a PhD in chemical education. When she first went to Knoxville she was a visiting instructor teaching laboratory sections and recitations. When she decided to do her PhD there she was able to get a fellowship. Her thesis was on the concept of cerebral hemisphericity, the idea that one half of your cerebrum processes information one way and the other half processes it another way. The left hemisphere is more linear, logical, sequential, and organized, while the right hemisphere is more holistic, not time-dependent, more creative, and less structured. Saundra wondered if students who were left-brain dominant wanted to solve problems in an algorithmic manner, using formulas; whereas students who were right-brain dominant would consider concepts to solve problems. For her thesis—"The Relationship between Cerebral Hemisphericity and Problem Solving Strategies Used by Selected High School Chemistry Students"—she measured students' cerebral hemisphericity using two physiological tests and a questionnaire. The physiological tests were dichotic listening (determining if the students heard a sound on the left ear or right ear while wearing head phones) and tachistoscopic viewing.[4] The questionnaire included questions such as: Do you prefer to use a map to get from one place to another, or would you prefer someone to give you stepwise directions?

Saundra wanted to see if there was a relationship between test results and the students' problem-solving strategies. She found there really was not a relationship. She also wanted to see if students using one method or the other were more successful at solving problems. She found that students who could switch back and forth between the two methods were the most successful problem solvers. If they tried an equation that didn't work, they used trial and error to be able to get the results. see what would happen. This suggests that more than one strategy is required to complete a task successfully. Saundra also tested for differences between race and gender but found none. She found students for her study by visiting local high schools and asking chemistry students to volunteer. She feels she had a good group of fifteen to twenty students, even though they were volunteers. Since it was a small sample, the study was called an exploratory investigation.

Her thesis advisor, Dr. A. Paul Wishart, was very supportive. He wanted to know what she was interested in doing, and she told him. Therefore, she was able to break the mold and work independently on her project instead of on the group project. He also gave her advice when she was putting together her advisory committee. He advised her to avoid potential members who did not get along with one another. He wanted her to avoid their infighting during her dissertation defense.

Saundra received her PhD in March of 1983, seven months after her husband left Oak Ridge National Laboratory to take a faculty position at Alabama A&M University. She arranged a position there teaching biochemistry and general chemistry. While teaching at Alabama A&M, Saundra did a lot of outreach and wrote grants to get funding for professional development programs for high school and elementary science teachers. The college was trying to help teachers engage students much more actively in the learning process. Saundra conducted teacher workshops presenting methods to make science classes much more active using experiments with everyday supplies, like baking powder, vinegar, washing powder, bleach, and so on. She was at Alabama A&M from 1982 to 1988.

Even though Saundra had a PhD when she worked at Alabama A&M, was in a tenure track position, and probably would have received tenure if she stayed, she preferred to follow her husband when he accepted a position at Cornell. She likes to say that they had a one career family— her husband had the career and she would get a job wherever his career took them. She decided to do this because she had wanted to follow her mother's example and not work outside the home when her children were young. Although, that didn't quite happen, she did work only part time until her youngest daughter started high school.

So, when her husband accepted a position as a visiting professor and they moved back to Cornell, she was not interested in a full-time position because her younger daughter was going into ninth grade. She rejoined the Learning Skills Center on a 75-percent-time appointment as an assistant director and chemistry lecturer. There, she conducted the weekly review sessions in organic and general chemistry and had some administrative responsibilities. She worked with the chemistry department while Nobel Laureate Roald Hoffmann and Professor Bruce Ganem were on the faculty there. Upon her return to Cornell she found many of the students who had gone to prep schools had a decent preparation for the rigors of Cornell, but many of those who had gone to inner-city, public high schools, though very bright, had not had the rigorous educational experience needed to be competitive at Cornell. To address the discrepancy, Cornell had a pre-freshman summer program for one hundred and seventy-five of the less prepared students. During that program, students were introduced to effective learning strategies and enrolled in a rigorous university course to help them with the transition from high school.

Saundra and her husband were at Cornell for twelve years, returning to Baton Rouge in 1999. Her husband had accepted a position as professor and chair of the Physics Department at Southern University, where he remained as department chair until 2010. He is now the James and Ruth Smith Endowed Professor of Physics at Southern University, Principal Investigator to the LIGO Scientific Collaboration (LSC) and Director of the Southern University-LIGO Advanced Optical Materials Laboratory.

Ironically, Saundra obtained a position at LSU as the director of the university learning center, the Center for Academic Success. She thinks that it is interesting she is at LSU considering the history of the place. After she'd been at LSU for three years, she was offered an adjunct faculty position in chemistry. She was later offered a position as professor without tenure. She readily accepted because she had no intention of going through the tenure process at that stage of her career. Before retiring from LSU, she became assistant vice chancellor for learning and teaching.

When Saundra first got to LSU, she thought she was going to do the same things she had done in Cornell as the director of the learning center. At Cornell, she did not work with the general student population and had an associate director to teach study skills one-on-one. When she got to LSU, they wanted her to interact with students at workshops and talk to individual students about how to improve their learning. She had never done anything like that before, and the resident expert, Sarah Baird, was on maternity leave. Saundra called Sarah and asked her to come back and work for 10 percent time—four hours a week. The first thing Saundra wanted was a demonstration, and she asked Sarah to do a mock individual session with a student while Saundra observed. After that, Saundra saw she could be very effective in helping students, but she needed to become familiar with a lot of new information.

She started studying and reading everything she could get her hands on and discovered the concept of metacognition—thinking about thinking— and knew she could teach students to think about their own thinking. When she explains this to students she likens the process to having a big brain outside your head watching what the brain inside your head is doing and then asking questions, such as, Do you really understand this information or are you just memorizing it for a test or a quiz? Anyone can be taught how to monitor and control their own thought process. This and other learning strategies can be very helpful and improve learning. For example, if one previews before class the information that will be covered in class, it will be easier to follow the class session. Students can pay more attention to the instructor's presentation because they are familiar with the territory, so to speak. They could also devise questions that they want answered. Most students just go in to a lecture and take notes, sometimes losing big picture concepts in the details, or vice versa. After Saundra learned these strategies, she started teaching them and found they gave remarkable results. She now travels to universities and conferences conducting presentations and faculty development workshops on how to use metacognition to teach others how to learn, speaking as far away as the University of Cape Town, South Africa, and Universidad El Bosque in Bogota, Colombia. She thinks one of the reasons that she gets invited to do so many talks is that many institutions want a science faculty member to make the presentation. Science faculty tend to believe strategies presented by non-scientists will not work for science students. They resist anyone whose expertise is in education or social science thinking, "It won't work in science classes."

Saundra feels she has had a blessed life, because she has been in the right place at the right time. When she started studying how to help students learn, the academic community was just beginning to be interested in improving student learning. Now institutions are being required to show that students meet specific learning outcomes. She is grateful that she was involved in the early days of developing strategies to help students improve their learning.

Saundra has trouble speaking about her career path, because she thinks she never had one. She was always focused on home and family and never really charted her career. She got a job wherever her husband's career took him and the family. But having said that, she thinks her job path has been very smooth. For example, when she received her PhD during the winter term, in March 1983, she wound up with an academic position beginning the same month, at Alabama A&M University. She was hired to teach biochemistry and two general chemistry courses because the professor who had been teaching the classes left for spring break and did not come back. Everywhere she has gone she has been able to a find a position doing what she wanted to do. However, she makes it clear, when she speaks to young women in the sciences, that her experience is not typical. Both while they are working on their PhD and when they have a traditional career, they will have demands placed on them and metrics to meet. This is something she never had to do.

Saundra was the first African American chemistry professor to be on the board of the Committee on the Advancement of Women in Chemistry (COACh), a group of senior women faculty that helps women successfully build careers in academia. She initially resisted as she was not a research-active chemistry professor and did not want to take the spot that another woman should have. But she relented, and when she joined the board expressed the concern that there should be other African American women on the board. She offered several candidates, supplying the names of Drs. Cornelia Gilliard, Sharon Neal, Joan Frye, and Gilda Barabino, all of whom accepted invitations to join the board.

Saundra values the insight and advice and perspective that mentors can provide. She has always listened to their advice and not been too defensive. She finds that some young people don't really want to take advice from anyone. They think that they can do it their way, so they make mistakes. She feels that mentors have had experience that protégés have not. Considering those experiences can help others avoid some of the pitfalls. She is grateful for all the people in her life who have helped. During her tenure as a student Dr. Wishart and Dr. Jack Howard Jefferson were her mentors. In March 2010, she organized a ninety-fifth birthday symposium for Dr. Jefferson at the NOBCChE regional meeting in Baton Rouge. Saundra first became associated with NOBCChE in 1986 and she thinks it is a wonderful organization. Her other mentors were Dr. Vandon White and Dr. William Moore at Southern University. She conducted undergraduate research with Dr. Moore. She adds that her husband, Dr. Stephen C. McGuire, has been a mentor since their days as undergraduates at Southern University. At LSU, she had several mentors in the field of administration.

Saundra is at the end of her university career but not her lifetime career. She will do a lot more speaking at universities and other institutions. She will accept more invitations so that she will be able to help teach faculty how to teach student how to learn. She now has a limited liability company called SYM Educational Consulting. She no longer thinks, as many people do, that some students are smart and therefore successful, while others are not. She thinks that all students who are successful have learned and implemented strategies, which led to their success. Less successful students have not learned the strategies for success. She now teaches Bloom's Taxonomy (Bloom's), which addresses the different levels of learning. She helps students reflect on the difference between studying and learning. One of her students described the difference this way: studying is memorizing information for a test while learning is knowing it well enough to apply the information. She also recommends that students work hard enough so they are able to teach the information, which means they have to work smarter. After she gets these concepts over to the students, she teaches them about Bloom's and specific strategies for working productively when learning new information.

Although Saundra works mostly with undergraduate students, she has also worked with graduate students who have been good memorizers in undergraduate school. She says she enjoys teaching the graduate students, because they need to be able to use critical thinking skills to succeed in their chosen profession.

When asked about her family, Saundra glows. She has two daughters she is very proud of. Her older daughter, Dr. Carla McGuire Davis, was born in 1972 and is currently a clinical associate professor of pediatric allergy and immunology at Baylor College of Medicine. Carla went to Howard University as a chemical engineering student and then to Duke University School of Medicine. She and her husband are the parents of Saundra and Steve's four grandchildren—Joshua, Ruth, Daniel, and Joseph Davis. Carla has a very active career and mentors a lot of students while managing her own children. Her husband, Eric J. Davis, is an attorney who helped start the Public Defender's Office in Houston, TX.

Her younger daughter, Dr. Stephanie McGuire, was born in 1976; she is an opera singer in Berlin, Germany, a mezzo-soprano, and has sung with the New York City Opera, the Boston Pops, and several other companies. Stephanie attended Howard University for one year before transferring to Massachusetts Institute of Technology (MIT) as a biology major. She graduated from MIT with an A average and was named a Marshall Scholar. She attended Oxford University and earned a master's degree in neuroscience and PhD in psychoacoustics. Then she went to Longy School of Music in Cambridge, MA, and earned both a master of music and a graduate diploma in opera performance. She received the Victor Rosenbaum Medal, given to the graduate who has achieved the highest level of both academic and artistic achievement, when she graduated from Longy.

What does Saundra think her greatest contribution to the field of science is? She thinks it was to bring together two separate silos of learning: the learning center community and the scientific community. The science faculty did not know that there was a whole discipline concerned with how to help students learn how to learn. For example, many science faulty do not realize that College Reading is an academic specialty focused on teaching students critical reading skills.

She disagrees with the idea that students who are not college ready cannot succeed in college. In fact, most students coming from high school are not college ready. No one has taught them the strategies necessary for college success. She sees the emphasis on college readiness as a way to keep some students out of college. She finds that students can become college ready while they are in college, if they are taught effective learning strategies.

When she talks to young women about their career, she tells them that they can't have it all. It is easier to mix work and home duties when the kids are in school than when they are younger and need more attention. Saundra feels women have to decide where they are going to deploy the resources they have and at what times.

Saundra is a strong advocate for HBCUs, because they are very effective at educating students, many of whom are unprepared for college, for success in graduate school, professional school, and careers. She says that if she had not gone to Southern University, she would never have had a career doing anything remotely related to chemistry, let alone in any kind of science or technical area. She needed the mentoring and encouragement she got from her faculty at Southern. By going to Southern where the classes were smaller and the faculty was nurturing, she absorbed their strong belief in students' abilities to excel. She believes faculty at HBCUs work hard to close the gap between where the students are when they come in and where they must be when the leave.

Saundra's hobbies are traveling, playing tennis, being outdoors, hiking, camping, and family things. She doesn't get to do much leisure travel, but she and her husband try to go on trips at least once or twice a year, to go fishing, swimming, and camping. When the girls were younger, they went with too.

What should students of the future know? Saundra says, "Recognize that success in any area is not going to be easy, but anything that is worth pursuing is worth spending a lot of energy and effort on. Intelligence is not static. The IQ that you may have in third grade can be higher in the tenth grade. But you must believe that." This concept that intelligence can grow reflects recent work in the area of brain plasticity. She wants young people to strive for 100 percent mastery. There are going to be what seem to be obstacles in their way. Maybe they are not real obstacles, only stumbling blocks. "Obstacles can be stepping stones to success. Find people who are going to be helpful to you. If you listen to positive people who know you can be successful, you will be successful. Mentor others. Share how you became the success that you are today."[5]

3.4 Sharon L. Neal

FIGURE 3.4. Sharon L. Neal

Sharon Neal (Fig. 3.4) is Associate Professor of Chemistry and Biochemistry and former rotation Project Director at the National Science Foundation.

Sharon was born on August 12, 1958, in Montgomery, AL, to Catherine Bell Neal and Alonza James Neal. Sharon arrived late in her parents' marriage, and by the time she was three, their marriage had dissolved. Her mother moved to find work in Philadelphia, PA, where her step-mother and niece lived. Sharon stayed in Montgomery with other relatives. Her father passed away two years later in January 1964, and she had little connection with the members of his side of the family after that. Her family life revolved around her sister and brother-in-law, who lived in Connecticut and New Jersey, and her mother's family in Alabama, where she spent most summers.

Education was important on both sides of Sharon's family. Her mother graduated in 1939 from Alabama State University, which was then called the State Teachers College, Alabama State College for Negroes. She earned a master's degree there in 1944. Mrs. Neal taught English and history in many schools across the state of Alabama. At that time in the south, it was common for educated women to travel to counties far away from home in order to teach. Later, she began to work in adult education and took additional courses at Temple University in Philadelphia. She may or may not have completed a second master's degree. Sharon's father was a tailor, who did not finish college, but he had two sisters who did: Catherine and Leola. Leola died young, shortly after she finished college. Sharon's Aunt Catherine went on to earn a master's degree and had risen to become assistant superintendent of schools in Montgomery by the time she retired.

When it was time for Sharon to go to school, she joined her mother in Philadelphia, where she attended primary and secondary school. Because she was a teacher, Sharon's mother believed it was important that Sharon get into integrated schools. and so enrolled her at Stephen Girard Elementary School, a public school in South Philadelphia, far from their North Philadelphia home. At first, the school put Sharon in Room 104, an integrated class. When the teacher realized that she knew both her letters and numbers and could read and write a little, they sent her to Room 101 with Ms. DeMarco. She was the only black child in that class. Because she lived far away, Sharon had to take the subway to school every day. At the start of the year, she went with an older girl who lived in the same apartment building. The girl really did not want the company of a little kid and was neither a good babysitter nor companion. Since Sharon was more afraid of the girl than of being alone, one day she decided to come home by herself. After that, she continued to make the trip alone, until she graduated in sixth grade. Her mom or sister may have watched her to make sure she knew what she was doing, but they didn't tell her. The hardest part for her was remembering to buy tokens for the bus and subway the one day a week they were available to students in the principal's office.

This was good preparation for Sharon, because in the seventh grade she transferred to the Baldwin School in Bryn Mawr, PA, a suburb more than ten miles away from her home. To get to school, she took the subway into the city and a commuter train to Bryn Mawr, catching the Paoli Local on the Thorndale Line Regional Rail every morning at 7:42—the years of commuting by subway had prepared her, and she commuted every day until she graduated from high school. Baldwin was an independent, college preparatory school. It provided a great education but was expensive. In the early years, her mother paid her tuition and expenses. When Sharon did well, she was able to get scholarships, which were a great help to the family.

Sharon does not remember learning much science in elementary school, but she remembers struggling with math. This is ironic given the role of mathematics in her PhD and current research. She got over her struggles with math in elementary school with the help of tutors. Her most important tutor was Mary Louise Windley, a young teacher who worked with Sharon's mother. Mary taught Sharon to attack math problems logically. Mary worked with Sharon for several years and eventually became a close family friend.

There was a full science curriculum in both middle and high school at Baldwin. In eighth grade, Sharon studied physical science with Mrs. Chesick and remembers the beautiful colors of the metal flame tests Mrs. Chesick showed the class, especially her explanation that the yellow color in the gas flames on the stove come from the salt in our food. Mrs. Chesick also taught physics in senior year. But physics was taught at the same time as Latin studies, including Virgil, the Roman poet, and Sharon chose to take Latin not physics, thinking she would have more opportunities in the future to take physics than to study Virgil. Besides, at that point, she did not expect to major in science, so physics did not seem critical. Sharon enjoyed Virgil and won city-wide awards in Latin. She now wonders if it would have been wiser to have taken physics and made Latin an independent study course. She did not remember if this option was suggested to her.

In high school, Sharon's chemistry teacher was Mrs. Arlene Dahl, a no-nonsense teacher, but with a warmth that made her approachable. The thing that really changed Sharon's experience in chemistry was getting sick in the fall of her junior year. She had lots of homework, which backed up while she was ill. In order to catch up, she had to spend a lot of time focusing on the stuff that she missed. While doing lots of stoichiometry problems, a light bulb came on and chemical reactions began to make sense. After that she didn't have a problem with chemistry, began to really enjoy it, and took advanced chemistry as a senior. In that class, Mrs. Dahl talked about molecular structure and properties. Sharon remembers her talking about carbon—the allotropes of carbon— and comparing pictures of the sp^2 hybridized carbon atoms in graphite to the sp^3 atoms in diamond. The teacher explained how those structures determined the slipperiness of graphite sheets, that could rub across each other, and the hardness of diamond. She explained that this in turn made graphite a good lubricant or pencil lead and diamond a good drill bit. Learning this made Sharon happy because she saw that the universe wasn't a completely random place; the world is not completely chaotic on the molecular level. This world made sense to her. She never expected to continue to have experiences like that in science, but she did realize chemistry and science were something she should stick with and study in college.

Since education was important on both sides her family, there was never any question about whether she was going to college. The only question was where. There were several factors that influenced the final decision. Sharon did not know if she should go to a HBCU or a predominantly white institution (PWI). Baldwin had prepared her to go to elite, competitive PWIs, but the feeling of being an outsider and having to speak and perform for all black people was not something she wanted to continue. She had seen an article about Spelman College, in Atlanta, GA, when she was a young girl and loved it. She started to say she was going to Spelman when she was in middle school. But when she did well on the standardized tests and started to get brochures in the mail from colleges like Stanford and Harvard University, she questioned that decision. But in the end, Sharon decided to go south because that was like home to her.

She applied to Spelman and was granted early admission with full tuition, but not room and board. She also applied to Duke University and was admitted with a full scholarship, including room and board and some work study. She thought hard about going to Duke, but in the end, she went to Spelman and is glad she did. She is sure that if had she not gone to Spelman, she would not have become a scientist.

Another factor in her decision was concern that a women's college (such as Spelman) was the wrong choice after attending an all-girls middle and high schools. She learned that the other schools in the Atlanta University Center (AUC)—Morehouse, Morris Brown, Clark, and Atlanta University—were close enough for Spelman students to take courses there. She didn't know it, but at that time, a number of the science courses were only offered at other campuses, Spelman's chemistry department was new, and its physics department had not been established yet.

When Sharon got to Spelman, she resisted majoring in a science because she thought all scientists were nerdy and isolated. Sharon's world was full of vibrant, engaged black women who seemed worlds away from her idea of scientists. Her mother was very socially and politically active. Sharon wanted to be social, wear pretty clothes, and have fun. She thought about majoring in economics because everyone in a bank looks nice, like they are part of the human race. But before long, Sharon had learned to look beyond the stereotypes of scientists and to accept her own inner nerd. She still liked pretty clothes, dancing at parties, and having fun, but embraced her studies and eventually decided to major in chemistry.

Another factor that helped Sharon become a chemistry major was the general chemistry course she took as a freshman. The course for non-majors at Spelman was a self-paced course, divided into modules. Sharon finished the modules before Thanksgiving and continued to go to class to tutor and help other students. This led her to more interaction with the faculty and she grew as a tutor. Because she found tutoring rewarding and remembered the stories of students her mother and uncles had helped, she began to think that maybe she could become a faculty member teaching chemistry. To do so, she needed to switch to the courses for chemistry majors. Since she had taken the non-majors course in the first semester, there was concern she was not prepared for the second-semester major course. To avoid falling behind, she decided to take an oral exam with Professor Henry C. McBay to complete the first semester majors course. Dr. McBay was famous for failing students across all the campuses of the AUC. Happily, her preparation with her high school chemistry teacher, Ms. Dahl, was solid, and he did not ask Sharon anything that she could not answer. She passed and became a chemistry major.

Once Sharon became a chemistry major she took courses at the AUC: in organic chemistry at Spelman with Professor George Sanzero; physical chemistry at Morehouse, with Professor Troy Story; and advanced organic chemistry with Professor Alfred Spriggs at Clark. She went back to Spelman for advanced inorganic chemistry with Professor Gregory Grant. It was fun, she made friends with students at all the campuses, but she did have to adjust to some students' competitiveness. In the second semester of general chemistry at Morehouse, she made friends with a guy who became upset when she had a high score on a lab report. He did not tell her what he got, but when he saw she got a ninety-eight, he stopped speaking to her. Organic chemistry with Dr. Sanzero was the opposite. He was a young faculty member, who had just earned his degree from Emory University. Like many young faculty, he was very idealistic and provided a very rigorous course. Sharon and several classmates bonded and did very well in the class. Later that year, a recruiter from Cornell University came to Spelman's chemistry department to meet students. It was too early for Sharon to apply to graduate school, but the recruiter was impressed by her enthusiasm for organic chemistry. He asked her to become a summer intern at Cornell. She accepted and worked in the lab of Professor William L. Dills in Cornell's biochemistry department, on the synthesis of alternative sugars. The Dills group was studying the metabolism of lyxitols and xylitols as alternatives to fructose- and sucrose-based sugars. They wanted to test the suitability for diabetics of sugars from those families. Sharon worked with a bachelor's level laboratory technician, Thomas R. Covey, and several graduate students. Tom taught her how to conduct many of the procedures they used to synthesize the sugars and supervised her efforts. He encouraged her to think independently. When Sharon ran one of the reactions three or more times without getting the expected product, she wanted to double check the method. The citation that they were using was in a paper in a journal in the library, but it was written in French. Sharon had studied French for five or six years at Baldwin, so she got a copy of the original paper and read it—she found that the glacial acetic acid they had been using was supposed to be ice water. In a later step, she almost ignited her hair with the green fire of a Raney nickel reduction. The building had to be evacuated when she used ethanethiol, which really stinks, in another reaction. Happily, she did not hurt herself, or anyone else, and learned a lot working with Tom and the rest of the group. Because of her correction of that mistake, Professor Dills put her name on the paper when he published the work a few years later.

Sharon's had another undergraduate summer internship at Lawrence Livermore National Laboratory, in California, where she worked with Dr. Dan Stuermer, a geochemist studying molecules released into the ocean by the natural oil seeps in Prudhoe Bay, AK. He had characterized a number of the molecules dissolved in ocean water using gas chromatography but a number of compounds hadn't been identified. When the lab group talked about it, they thought the unidentified compounds might contain sulfur. Sharon pursued this idea, purchased small standard amounts of several sulfur compounds, and tested them using gas chromatography. Several of the standard compounds had the same retention time as an unidentified compound. When the samples were investigated by gas chromatography–mass spectrometry (GC–MS), the data confirmed the compound identities. Sharon was listed as a co-author on that publication as well.

Sharon loved living in Atlanta, and Professor Sanzero spoke highly of his training at Emory University, so she decided to apply to Emory for graduate school. She was accepted and awarded a teaching fellowship and went to work with Dr. Fredric M. Menger, who was well known for his work on surfactants. He built molecules that had unusual structures and interesting properties driven by the hydrophobic effect. Sharon thought the work was great, but the two of them did not communicate well. She was frustrated by the tedious work he assigned her but was not able to talk with him about it effectively. Now Sharon thinks he was asking her to do fundamental work to prove her competence, but at the time she did not recognize this concern. She thought he did not want her as a student and was giving her throw-away work. She believes that communication about expectations and competence between white faculty and students of color, especially women students and students from HBCUs, are fraught with this kind of misunderstanding and require more patience than either party typically has.

Rather than continue to be frustrated, Sharon took a leave of absence from Emory and got a job at the Coca-Cola Company, in the Corporate Quality Assurance Special Project Unit, which she enjoyed, because she learned how to do many types of chemical analyses—she joked that she learned "every analytical technique known to man." The unit's main task was to monitor ingredient quality of samples that could not be imported into other countries from the United States because of trade laws. These laws would not permit some ingredients to be imported without proof of their quality. However, the unit also investigated unusual properties in ingredients or products. For example, when a weird taste was discovered in the Sprite syrup from a plant in France, staff at the unit figured out that an O-ring had fallen into a mixing vat. Soda is made in large vats; and possibly something happened to cause the O-ring to fall into the vat and change the taste. Even though she enjoyed the job, it was clear that it was a dead-end, with no path for advancement.

Around the same time that Sharon was realizing that there would be no advancement for her at Coca-Cola, Professor Isiah M. Warner moved to Emory from Texas A&M University. She went to his job talk and met him and his family at a social event. He looked up her file, saw her grades, scores, and the things she had done and decided to recruit her for his group. He took her to lunch and told her that she should be in graduate school. He also said he thought his group was doing some things that she would like. He invited her to see his lab, meet his group, and see what his students were doing. Sharon met the group and decided to re-enroll at Emory in Dr. Warner's lab. Dr. Warner was working to develop more informative fluorescence measurements. Fluorescence measurements record the intensity of light emitted by certain molecules. As a graduate student, Professor Warner had used a video camera to record the intensity of light emitted versus the color of light absorbed and the color of light emitted at the same time. The video camera measurement produced an image that maps a sample more thoroughly. It produced a two-dimensional data set—a data table—that reflects the sample composition. He had written software that would resolve a data table of a two-component mixture into the data tables of the individual components. Now, he wanted to generalize the method and increase the number of components that could be studied. Sharon was the first person to generalize the problem, writing software that resolved fluorescence data tables containing any number of components. For her degree, she demonstrated the resolution of mixtures containing up to five components. In principle, it does not matter how many components the mixture has, if the component produces a distinct detector signal, it can be discovered. Of course, most real measurements have shortcomings that put limits on that. It was challenging, but Sharon loved working in the new field emerging at the interface of applied mathematics and analytical chemistry. The new field was being called chemometrics. Chemometrics is the use of mathematical and statistical methods to improve the understanding of chemical information and to correlate quality parameters or physical properties to analytical instrument data.

Sharon did not often interact with the other graduate students in Warner's group on science projects, since she was the only one doing chemometrics. But she was friendly with several of the other women in the group. In her last year, her former roommate, Dr. Denise Sylvester, was a postdoc in the lab. Overall, Sharon's strategy for getting through graduate school was to put her head down, put the blinders on, and focus. Her idea was to "go straight forward, don't look to the left, don't look to the right and don't get involved with laboratory politics." Now Sharon realizes that this may have been a good short-term strategy, but thinks it is unwise in the long run. A faculty member is at the hub of all the laboratory politics, and students need to gain experience in that role. This is especially true for women and students of color.

Toward the end of her graduate studies, Dr. Warner sent Sharon to study with Professor Ernest R. Davidson, with whom he had studied while a graduate student at the University of Washington. By this time, Professor Davidson had moved to Indiana University, so Professor Warner sent Sharon to spend a summer there. Dr. Davidson is a very eminent quantum chemist and a member of the National Academy of Sciences. He taught Sharon a lot about formulating mathematical problems and writing more efficient software code. He introduced her to BLAS (Basic Linear Algebra Subprograms), which is a library of highly optimized computer routines that programmers use to build more efficient programs. Before that, Sharon had been writing FORTRAN code, which was inefficient and hard to debug. She learned so much from Dr. Davidson that he is a co-author on the paper on her graduate work, which was published in *Analytical Chemistry*. She also wrote a few other papers on smaller projects at that time.

Isaiah Warner was not only her PhD advisor, but also one of her mentors. Another mentor was Dr. Henry Blount, who worked at the NSF. When he first visited Dr. Warner's lab at Emory University, Sharon was asked to show him her work. Since she was busy at the time, she grudgingly showed him her results. Happily, her enthusiasm made a positive impression on him. He gave her good career advice and later when she was on faculty, recommended that she receive her first NSF award. Another good friend of Dr. Warner's who became a mentor to Sharon was Professor Linda B. McGown, a faculty member at Duke University when Sharon was at Emory. She and Dr. Warner both earned their PhD degrees under the supervision of Professor Gary D. Christian at the University of Washington. Sharon found Dr. McGown to be a good source of advice and she wishes she had sought her out more often. Sharon did not apply for a postdoc with her because she felt McGown's and Warner's research areas were too similar. Since Sharon was the only one in Warner's group doing chemometrics, she decided it would be better to work with a group with that as a focus. Later, she wondered if close guidance of an effective woman advisor would have been of more value than scientific breadth.

She received her PhD and went as a postdoc with Professor Bruce Kowalski at the University of Washington (UW) in Seattle to connect with the chemometrics community. She found it as difficult to communicate with him as it had been with her first thesis advisor. During the postdoc, she tried to write software to resolve Nuclear Magnetic Resonance (NMR) data as she had for florescence. She thought NMR was a good contrast to fluorescence. NMR detection limits are not as good but have much higher information content. Unfortunately, most NMR data sets are huge, so the problem of computer capacity and speed were significant at that time. Sharon ended up spending most of her time figuring out how to get calculations to run, instead of thinking about how to use the data to understand the sample. She made a little progress using graph theory ideas, but never really got where she wanted with that project. Later, Sharon concluded it would have helped if she had focused her thinking and discussions with others on figuring out how to break this problem in to smaller steps. At the time, she did not see that. Since she and her advisor were not communicating very well, the postdoc ended at that point.

After her postdoc, Sharon returned to Spelman to teach, which she enjoyed, because she loves interacting with the students. She found it easy to provide rigor in her teaching, as Drs. McBay and Sanzero had done for her as an undergrad. She wanted to give Spelman students really high-quality education, but not scare them. Over time she learned she must convey to students that her efforts to be rigorous are motivated by her desire to see them do well and have every opportunity. When she can convey that to students, classes go much better. If they feel that she is making the material hard just to show how smart she is, students can become hostile. She found it was easier to convey to young black women that her efforts come from concern. She knew many did understand what she was trying to do. By the time she left Spelman, she felt she was connecting well with the students.

Sharon was also able to do some research at Spelman. She worked on software development and in the summer was recruited to participate in a US Naval Research Lab program for HBCU faculty and worked with a group that did molecular modeling. It was very interesting and fun, but not very productive, as there was not enough time to complete the work over the summer months. Overall, she found it difficult to make the regular work she was doing fit into the undergraduate framework of taking the summer off, so she eventually left Spelman, even though she was on the tenure track there. While still at Spelman she was invited to interview at the University of California at Riverside (UCR). Her plan was to visit just and practice doing an interview. But when she was offered the tenure track position, she decided to take it. There, she used fluorescence to study lipids, which form cell walls in many animals and plants. She assembled a new instrument and wrote software that recovered a lot of information about the samples using methods similar to those she had used in grad school. However, the fluorescence of molecules in model cell walls was complex and required an extra step to sort it out completely. She had to choose whether she should try to figure out a way to do the extra step or find another system to study. She decided to stick with it and figure out how to do the extra step, but none of the solutions she came up with worked very well. She ended up publishing work on simpler systems, but too much time had passed and the faculty in the department did not think what she had done was enough. Now she says, in her youthful inexperience and idealism, she framed that choice naïvely. Rather than seeing the choice to put off solving the extra step as avoiding a hard task, it also could have been seen as taking the time to build more resources to tackle the hard problem better. If she could have generated more publications and raised more grant money, she might have had more students to work on the complex system. On the other hand, it may not have mattered.

That was not Sharon's only naïve choice. She was intense and idealistic and thought being very hard working would compensate for any mistakes. She was the first woman and the first black person in that department. She knew communicating with the other faculty would be hard, but she did not have a clear plan for making it work. She did not really try to form relationships with other faculty that would produce meaningful discussions and understanding. The senior faculty did not seem to think anything more was needed beyond giving her start-up money and lab space. She thinks their concern about her mistakes grew to mistrust. Buyer's remorse about hiring her set in.

She was invited to give a talk and apply for a job at the University of Delaware (UD). She got the position, also tenure track, and took it. In addition to university start-up funds, the NSF grant she'd received at UCR was renewed, which allowed her to set up a new lab and do new experiments. She worked with several students doing more research on lipids and got tenure in 2002. Sharon never had a large research group, but the challenging thing at UD was teaching. It can be hard for young women to be taken seriously when they teach hard science classes. In her case, the problem was a big gap between the amount of work in the class and the credit students earned. The class had been taught with this imbalance for years and though the students didn't like it, they had accepted it. However, when Sharon taught the course, using the same syllabus and textbook as the other faculty, it was like all hell had broken loose because the students were furious. The department administration did not understand the dynamics, so they did not express confidence in Sharon when students complained and she received bad teaching evaluations filled with anger and insults. Sharon handled this by deciding there was a lesson in it for her—that focusing on student complaints about her lack of competence was unproductive, and ironically, led her to make the material harder and undermine student understanding. So, she worked harder to make the material relevant and clear so that students would know what they needed to do to and to reassure students that she wanted to see them advance. When she talked to students she said to them, "You can do this. And when you have a failure, you can get up. You can keep going."

While living in California, Sharon was involved with a neuroscientist, John H. Ashe. When he became ill in 2004, Sharon took a medical leave and went to California to support him and was with him when he passed away in October 2004. When she got back to UD, she requested a sabbatical. She found out about a position from Dave Haaland, who worked at Sandia National Laboratories in chemometrics. He connected her with Margaret Werner (Maggie) Washburne, a geneticist at the University of New Mexico (UNM). It was clear to them that the computational methods that they were developing for spectral analysis could be used for microarray data, a type of genetic assay used to monitor expression of enzymes. Sharon was interested in working with Maggie to learn about the nexus between genetics and chemistry. She went to UNM in June of 2005 and worked there for a year. There, Sharon took Maggie's class in advanced genetics and Andreas Werner's class in computational biology. She thought the classes were fun. She also made many friends, including Deborah Hunka, then a postdoc at Sandia, who now works in chemical and biological warfare agent detection. They sat together in the computational biology class and became fast friends. Sharon says they were the two best educated people in the class and the worst behaved.

Maggie was doing network analysis to see how the various pathways of yeast gene expression were interacting with each other. Sharon developed some network software, but the genes that were being highlighted weren't the genes Maggie's group was working on. If she had been a graduate student she could have stayed and found a way to link her work to theirs, but there wasn't time for that in a year. It did not really go anywhere, but the work was very interesting.

When Sharon returned to Delaware from her sabbatical, she found that her mom could no longer live on her own. So, she purchased a home where they could live together. By connecting to the Delaware Division of Aging she was able to find adult day care for her mother. So, she took her mom to the senior center near campus at 8:00 AM, then went to work. Sharon needed help from caregivers because her mother had to be picked up by 4:30 PM, which was a short day for Sharon. It was challenging to find consistent, high quality help. They made it work for more than three years. Then in 2011, her mother fell and broke her hip. The surgery to repair the hip went well, but dementia had progressed and Mrs. Neal she was not able to regain her mobility. So, she went from the rehab hospital to a nursing home, and Sharon shifted from caregiving at home to caregiving at the nursing home.

Sharon tried to keep her research going, but knew she was falling behind. Since she was on the Math and Physical Sciences Directorate Advisory

Committee at the NSF, Sharon knew the NSF Chemistry Division needed program directors. She thought serving as a program officer would be good stimulation for her career. As she did not have any graduate students and her mother was in the nursing home, she could change direction. In 2011, she became a rotating program officer at the NSF Washington, DC. The first year she served as a program officer in the Chemical Structure, Dynamics and Mechanisms (CSDM) program, where the lead program officer was Dr. Colby A. Foss. CSDM is the program that supports most physical chemistry and some descriptive chemistry. The CSDM Program supports research projects that have strong implications for advancing the foundational knowledge of chemical systems. The Program supports research on the nature of chemical structure, chemical structure property studies, chemical dynamics, and chemical mechanisms. Sharon was responsible for selecting reviewers for grants submitted to CSDM and then compiling the reviews into analyses that would support a recommendation to fund a project or not. The program officer can reach a different decision than the reviewers, but that decision must be well supported and this does not often happen. Sharon thought it was great to be "behind the curtain" of science funding, demystifying the obscure process. She also liked the work very much because it was intellectually stimulating.

Sharon did not to move to Washington, DC, to take up the NSF position, but took advantage of their telework policy and commuted; for her mother to receive the care she needed in the nursing home, Sharon had to be there consistently. Although nurses feel sorry for people who don't have family, they are overworked, and, if they cut corners, they are more likely to do so with patients who are alone. So, Sharon commuted by train on days she needed to be in the office and worked from home the rest of the week. If she needed to stay late, she would stay in a hotel.

In the next two years, Sharon worked with Dr. Lin He and again with Dr. Colby Foss to fund analytical chemistry programs in the Chemical Measurements and Imaging program. She liked learning about the beautiful work scientists were doing. She also worked to advance data science. As she became more expert, the limitations that commuting put on her time got harder to manage. But, she was also concerned about her mother's health and did not think she could be moved to DC. She decided not to apply for a permanent NSF position.

When Sharon returned to UD, there was a new department chair who thought the work she had done at NSF was meaningful and helped her get university support to restart her research. He also suggested she collaborate with an environmental engineer on developing a model to predict the light-induced toxicity of compounds introduced to marine settings by oil spills. Sharon's group had done spectroscopy measurements to support building the model. They began doing photochemistry work inspired by the collaboration. Since the work they are doing now is more interesting to students, she has two undergraduates and four PhD students.

Sharon is also trying to develop a collaboration with an instrument company. She has an idea about how to make the equipment for taking fluorescence measurements more accessible to a larger audience and is hoping to make it work. Sharon is still working on numerical methods, but she is trying to be sure the motivation for it and the value of it are clearer. She has been working on making clearer connections to things that people understand, because tax-payer-funded research depends on those tax payers being able to see the potential benefits.

Sharon is a member of the ACS and has been a counselor and local section officer. When her mother's health declined she had to give up a lot of that work, and now mostly just pays her dues. But, she was asked by the program chair for the Analytical Division to organize a session at the national ACS meeting in Philadelphia in 2016. She was glad to do it, since it gave her the opportunity to hear from a diverse group of scientists. She says it shows, "We're here and we care about the science, just like everybody else." She also thought it was good to do it since the session included black women chemists, black women don't often play that role at ACS meetings.

Sharon is also a member of the COACh Advisory Board. This group develops and delivers workshops to help women develop the soft skills they need to advance their careers: negotiation, leadership, work/life balance, and so on. COACh started with a focus on chemists and chemical engineers but has expanded to include other fields. Mathematicians, geoscientists, physicists, and other scientists sit on the board, and COACh workshops are presented to women and men in many fields. Sharon has been on this board for about ten years. She thinks it has helped her to understand that although she has limitations and has made mistakes, many of the things she has experienced are not about her alone. Many woman, not simply women of color, in scientific settings have had similar struggles. This work with COACh has been critical to her perseverance.

Sharon is also working hard on service at UD. The graduate students want to establish a chapter of NOBCChE, and as one of the faculty advisors, Sharon has been helping them get started. She also works with the ADVANCE Institute at UD. Funded by a large NSF grant, the institute sponsors many programs and recommends changes to increase opportunities for women faculty. A signature program was a conference entitled "Women of Color in the Academe: What's Next?" co-sponsored with the Offices of the University President and Provost. Sharon and history professor Anne Boylan were co-chairs of the conference program committee. One of Sharon's favorite moments during the conference was the book signing event at which authors were able to briefly present their work to the audience and autograph books. It made Sharon cry to see these scholars recognized for their intellectual accomplishments. She was touched because she doesn't see women of color applauded for their scholarship and creativity very often.

Mrs. Neal died on April 27, 2015, and Sharon took her mother home, as she wanted, to be buried in Montgomery.

Sharon is single. It may be because her parent's marriage dissolved, but she thought she needed to be careful and not just marry to fit in. She loved John, her boyfriend in California, but when her position there didn't work out, she did not choose to give up being a scientist to stay with him. She thought she would get to a point of stability in her career that would allow them to live in the same place again, but then he became ill and died. She wondered about finding another man after her John died. Her life had not been as she had imagined when she was younger—getting married and having children. Later, she met Jesse Coleman, who worked at UD as a substance abuse counselor and they hit it off. He helped her for years with her mother and her house, so she is not alone. She is glad she met Jesse, because he has been very kind and supportive of her, especially when her mother was ill. Her Spelman sisters Melva Ware and Wendie Willis also helped her a lot before and during her mother's illness and passing. Rev. Claude A. Shuford, pastor of her family's church in Montgomery, was also very kind.

When asked what she thinks about the field of chemistry, she replied that it is hard to give a short answer. Chemistry is a really important field and It is important that chemistry is advancing into other fields. Chemists are studying the molecular roots of other fields. Chemistry is important in biology, geoscience, and astronomy. She also thinks advancements in computing are important in chemistry. Computer models of molecules are getting larger and more accurate, so there can be more synergism between experiment and computation. She thinks the reduction of the barriers between fields and sub-disciplines is great and looks forward to more cross-fertilization between fields.

When asked about science funding in the United States, she says our national politics have made the future of science funding uncertain. She worries that our politicians do not embrace science advancement as they should do if they know history. Basic science has been the engine producing technological and economic innovation. Therefore, it should be supported by everyone. Some people view science as a luxury. She thought that some US presidents were supportive of science but did not know if that would be true of all future administrations. Currently there is just not enough money to fund all the good ideas scientists have. She thinks that people who have the power of the vote should be informed about the importance of science. This means scientists need to be ready to share in understandable language the importance of science to the public.

When asked about the future of minorities in science she is upset when she hears people say minorities belong in sports, but not science. The idea that some think that particular groups can do the science and everybody else can stand around to watch is crazy. She said science is an essential human function. All humans are curious and we all want to understand our environment as a part of understanding who we are. We are not going to know who we are as a species when we have whole segments of the species excluded from essential human functions.

When asked about the education of young people, she is concerned that there is too little transfer of knowledge. Too often, students have been taught concepts, but when they get to a new class, they feel like they have never seen it before, because it might be taught differently. She is also concerned that as human knowledge increases, education becomes more superficial. For example, most students understand how to use a calculator, but few grasp the abstract relationships between quantities. She hopes more students will look to education to answer their questions and not be satisfied to simply get a grade.

What would she tell a young minority, future scientist? There is so much beauty in the universe. Part of being a scientist is the joy of being able to witness and appreciate that beauty. For many people, science is the work our souls must have. If that is you, hold on to it and do it. It is worth it. You must educate yourself, not just about science, but also your situation as a minority in whichever scientific setting you work in. If you do, you can preserve your ability to enjoy the work and benefit yourself and others from that enjoyment.[6]

3.5 Mandë Holford

FIGURE 3.5. Mandë Holford

Mandë Holford (Fig. 3.5) is a professor of biochemistry at Hunter College and has scientific appointments at the American Museum of Natural History (AMNH) and Weill Cornell Medical Center in New York City.

Mandë's mother Janice and her father Winston Holford are from the Co-operative Republic of Guyana, South America. In the United States, her mother was a supervising chef for an airplane catering company, LSG Sky Chef. Her father worked for the Metropolitan Transportation Authority (MTA), although he didn't drive a bus or a train. They were a middle-class family, who worked hard to ensure that their children would be able to enter college.

Mandë is the middle child of five; she has an older brother and sister and a younger brother and sister. She grew up in Brooklyn, NY, on Eastern Parkway in a building across the street from the Brooklyn Museum. Her mom and dad were great cooks, and there were always many friends who lived in the building who came to their meals.

Mandë went to Brooklyn PS316 elementary school. She was a very shy child who liked to read and could in fact, read very well for her age. But the placement exam was an oral exam. She was so shy, she failed and was placed in a special needs class. The teacher in that class soon discovered she was very good at reading. So, after three months, she skipped two grades and went into the third grade. In that class her teacher, Ms. Steele, recognized her potential and encouraged her love of reading. Since Ms. Steele lived on the same block, they would walk home from school together. Ms. Steele encouraged her to get ready for the gifted and talented program exam for high school. In sixth grade, Mandë took the exam, skipped seventh and eighth grades, and went directly into the ninth grade. There were three NYC high schools in the program at the time: Bronx High School of Science, Stuyvesant High School, and Brooklyn Technical High School (Tech). Since Mandë's family lived in Brooklyn and the other two high schools were quite a distance away, she had already decided she wanted to go to Tech. But her family moved to Rosedale in Queens, and she went to middle school there for one year. She then took and passed the exam to get into the gifted and talented program and go to Tech, which meant a commute back to Brooklyn from her new home. At thirteen she started tenth grade at Tech, catching a train every day at 6:00 AM to be at class by 8:00 AM. She was younger than her group of friends and interested in different things. A tomboy, she loved sports, and was on the track team in both elementary and high school. She competed in track meets around the city, accompanied by her dad who cheered her on.

At Tech, Mandë was a math and science institute major in the Noble house. (Tech was originally an all-boys school with the tradition that each major had their own group, called a house.) She thinks Tech was wonderful—the teachers were stellar and she had a great group of friends. She studied AP calculus, AP chemistry, and AP biology. Since there were big gaps in her education, she isn't clear about whether she had chemistry before AP chemistry. There was a gap in her arithmetic knowledge as well. She was accepted at several colleges, including Barnard College and Wesleyan University, but did not qualify for financial aid. She decided to go to Wesleyan University in Middletown, CT, because she got a merit scholarship. Barnard only had scholarships for students who had no other choice. Even with the scholarship, expenses were close to thirty thousand dollars when food, books, and so on were factored in. Her parents told her that education was important and her one job was to go to school and do well. On Labor Day Weekend, they packed her trunk into the car, drove her the campus, dropped her off, unpacked her as quickly as possible, and left her. They wanted to get home for the big Labor Day weekend in Brooklyn, which is capped by the Caribbean Day Parade.

Wesleyan is a liberal arts school and Mandë had a wonderful time there. She planned to be an international lawyer because she was very interested in global problems. She took German history and literature and other global subjects. But Mandë was unable to stay at Wesleyan. After only a year, the financial burden proved too much on her parents, especially considering the needs of her four other siblings. She returned home feeling her parents had broken their contract—that if she did well in school she could go to any college she wanted. She decided to go to City University of New York (CUNY) and "to spite her parents," she decided to pick what she thought was the worst CUNY school, York College in Queens. As it turned out that choice proved providential, because there she met Dr. Lawrence Johnson, who was working in the chemistry department. When she did well in his general chemistry course, he invited her to do summer undergraduate research in his lab. No such opportunity would have been available at Wesleyan.

She had no idea what working in the lab involved. In high school chemistry lab, the experiments had followed recipes with predetermined outcomes. At York, Mandë was exposed to "a lot of wonderful unknown stuff, like yttrium aluminum garnet (YAG), lasers, and burning holes into molecules." She could not believe that you could get paid for coming to work, asking questions, touching things, trying experiments, and breaking things! The National Institutes of Health (NIH) program Minority Biomedical Research Support (MBRS) gave her a stipend for doing research, and so she could tell her parents she had a summer job and she was getting paid to do research.

Mandë came to believe Dr. Johnson's claim that chemistry was a career—a good career—so, he was instrumental in putting her on the science path. She spent a total of three summers in his lab before graduating from York College. He also guided her to further study by saying if she wanted to be a scientist, she should go to graduate school. When considering graduate school, she wanted to get out of New York where all her education had taken place. She interviewed and was accepted at both Penn and Yale University. When she went to visit, their campuses were beautiful, but students she talked to did not seem very happy. Her last interview was at Rockefeller University on the East Side of Manhattan in New York City, where Dr. Johnson had encouraged her to apply. She went to the orientation and was amazed at the school. It looked like another world, with a gorgeous campus and palpable excitement for doing science. The students and the faculty smiled and everyone seemed happy! She learned that there was a lot of support, and funding, for research and students are treated as scientists from day one; there is a written pact—Rockefeller would train you to be a top-notch scientist. She decided she needed to go there, knowing that acceptance automatically meant working for a PhD. (Other schools require candidates take qualifying exams to enter a

PhD track.) When she told her parents she was going to Rockefeller they were devastated, thinking she meant Rockefeller Center with its famous music hall and ice rink. However, when they came to visit and saw the campus, facilities, and her apartment, they were pleased. Since her dad liked to cook and was worried about Mandë's cooking, he brought her lunch everyday throughout graduate school, getting to know the guards and everyone else.

At Rockefeller, there was no set curriculum, so Mandë could construct her own. The faculty was taking turns being the Dean of the college at the time, which contributed to the looser structure. She says Rockefeller is very good if you have initiative and are self-driven. As for coursework, there was a limited supply of classes because most of the faculty did not teach. Since Rockefeller is a biomedical research institute with the mission to benefit humankind, it was geared toward basic research looking for therapies to treat illness and disorders. Rockefeller did not have a lot of courses about the physical sciences, math, chemistry, or physics at the time. It was very heavy on biology, and biology courses, like cellular biology with Sandy Simon. However, students could take courses at Columbia University or other schools. Mandë took classes on Nuclear Magnetic Resonance Spectroscopy (NMR) with Art Palmer at Columbia. She also took advanced organic chemistry there, when she decided she needed it for her research. Rockefeller did not give credits or grades. One took the classes to gain knowledge. So Mandë does not have a traditional transcript. Rockefeller's simply lists the course titles and whether she passed or failed.

Most wonderful about Rockefeller, was that it made it easy for graduate students to focus on science. Rockefeller's philosophy is that faculty and students are there to do science. So, the university takes care of all the distractions. They provide housing, health care, and recreation—a tennis court, but no pool. When the student association griped about the lack of a swimming pool and wanted one put into a new building under construction, they got a new gym instead, a perfect illustration of the school's vision.

There is no tuition fee at Rockefeller. Mandë received a stipend of about twenty-six thousand dollars per year. The university subsidized housing, and Mandë thinks she lived in one of the best New York apartments she will ever live in. Her share of the apartment's subsidized rent of about two to four hundred dollars, was taken out of her stipend check. Her roommate, Linda Wilbrecht, worked in Fernando Nottebohm's lab studying bird songs, learning, and memory.

To encourage their students to be well-rounded scientists, Rockefeller maintains a great fund called the Detlev Bronk Development Fund, which gives students the means to go out and enjoy the city, for example, to buy tickets to Broadway plays, jazz performances, or other cultural events. Mandë used her money to buy season tickets to the New York Knicks (basketball team), because she is a sports fan.

Mandë studied at Rockefeller for five years and thinks the training she received was stellar. There was always energy and excitement to do science and that attitude has remained with her throughout her scientific career. There was no such thing as "no." If she came up with an idea, the first response would be, "Well, let's see how that might be possible." She said when you are trained in that kind of environment, you tend to think more broadly and more creatively about your science. You might want to try out a crazy idea to see if it will work, and the administration staff and others will say, "Yes, try it."

While she attended there, Rockefeller celebrated the hundredth anniversary of the graduate program on campus. During those hundred years about twenty Nobel Prize winners had worked there, which worked out to someone winning a Nobel about every five years. Nobel prize winners still on campus practiced an open-door policy. Anyone could speak with them.

During her time at the school, Mandë's class of PhD students numbered about twenty—out of a student body of about a hundred. There were more postdoctoral fellows on campus. But, there were only two minority women in the graduate program, Marsha Henderson and herself. Mandë never found any other minority women in any of the courses she took.

Mandë's thesis research was on chemical ligation strategies for increasing peptide synthesis to better study their structure and function and their activity in biological systems. She worked on making peptides longer than fifty amino acids, the traditional limit of solid phase peptide synthesis. To do this, she used a technique called expressed protein ligation (EPL), which combines chemistry with recombinant biology. Recombinant biology allows you to express (synthesize) a protein using bacteria, and then purify it. This is done by making a specific plasmid, which is a circular piece of DNA. You insert the DNA into bacteria, like *E. Coli* (*Escherichia coli*), grow it up in broth, and then purify out the specific protein. That's how large proteins were normally synthesized. Expressed protein ligation (EPL) allows the use of solid phase peptide synthesis to synthesize part of the protein, so the researcher can manipulate cross-linkers, fluorophores, radiolabels, or whatever they like and incorporate them at specific sites. Then, using a ligation technique based on native chemical litigation, reacting an N terminal cysteine with a C terminal thioester, it's possible to synthesize very large proteins with site specific probes in order to study the protein's structure and function. That's why the technique is called expressed protein ligation. The biologically made and chemically synthesized parts of the protein come together kind of like Velcro and make a natural peptide bond. With EPL most of the protein is made recombinantly, letting the bugs do the work, so to speak. The part that the researcher is interested in manipulating or studying is made synthetically. The parts come together through the EPL reaction to make both a native peptide bond and a native form of the protein of interest with a site-specific tag.

Mandë applied her Velcro technique to a lot of different proteins of various sizes to show the range of possibilities using the expressed protein ligation technique. She was exploring what a specific protein interacts with in the cell. For example, she would add a fluorophore to a protein to be able to trace it using fluorescent methods. Doing things like that is a powerful way to mechanistically understanding how a protein functions in the environment of a cell—any type of cell.

In a curious synchronicity, when her chemistry advisor Tom came to Rockefeller, he had to figure out what chemical tools would be important for biologists and their biological research on campus. EPL was one of those tools. So Mandë learned not only that chemistry can offer tools that are important for biologists, but also that interdisciplinary science is the wave of the future. It is most important to collaborate, to interact with people beyond the folks in your lab. Science doesn't have silos or shouldn't have. Tackling very big and important questions should be done collaboratively across disciplines. In her research, Mandë had two main collaborators, Seth A. Direst and John Kuriyan, both structural biologists. This principle of collaboration has guided her in her own postdoctoral research, when she was looking for a position, and currently when she does things in her independent laboratory.

Mandë was able to publish a lot of her PhD work, including an important publication with fellow graduate students Morgan Huse and John Kuriyan about the hyper-phosphorylation of the TGFβ receptor, which identified, for the first time, how that receptor worked. Other publications on her PhD research were more about developing methods, for example, how to make EPL work.

Because she is interested in other aspects of science, while working toward her PhD, Mandë went to the American Museum of Natural History (AMNH), contacted one of the curators there, told him that her PhD research was in proteins and peptide chemistry, and asked if there was any way she could contribute. Robert Deale, who was doing an exhibition on the genomic revolution, needed a model of a receptor in the eye, the rhodopsin protein in the retina. Mandë knew how to make that model using molecular visualization software called PyMOL and other programs! The exhibition highlighted personalized medicine and explored what it meant to solve the riddle of the human genome, what a genome is, what genes look like and what proteins are. Mandë worked on the exhibits and did a couple of things with protein microarrays, which, at the time, were becoming more important because of the human genome research. They were highlighting personalized medicine and what does it mean to solve the genome, what is a genome, what do genes look like, and what are proteins. She found it was a great learning experience.

Like most people, Mandë had not been aware that the AMNH and other science museums have active research labs and research staff. So, they not only do great science, but also help to communicate that science to a larger audience. They give back and help shape society through effective science education. Mandë was hooked. She wanted to bridge what she was doing at the museum with the biomedical research she was doing at Rockefeller. So, for her postdoctoral study she looked for a lab that could do those two things.

To complete her PhD she had to write a thesis proposal and dissertation, defend it before a committee, and then defend it before the public. While completing her PhD, Mandë read and was influenced by Bruce Merrifield's book *The Golden History of Peptide Chemistry*. She used quotes from this book about success and failure in each of her chapters and concluded with a quotation from his Nobel Prize publicity tour: "It has all been very interesting and rewarding, and the opportunity of a lifetime, but now it's time to return to reality and resume my life, because so many things have yet to be done." She loved that quote because it reflected her life at Rockefeller and had included it in her lab book thesis notes.

After the defense, she held a party on the thirty-eighth floor of Scholars Residence, a sort of dorm, if you can call the subsidized housing at Rockefeller dorms. It was a whole-campus, community party. Everyone from the dean's office, all her graduate students, and all her family came. She even invited all the guards that her dad had gotten to know. And was just wonderful! Her mom ordered three cases of champagne, stocked two refrigerators, and everyone drank all night to celebrate a PhD scientist daughter.

When Mandë left Rockefeller, she went to the AMNH for a year and worked in their education department, where she helped to develop an after-school science program called the High School Science Research Program (HSSRP). It was her way of trying to give back and engage urban students in exploring the excitement of science. As director, she guided after-school programs in genetics, astronomy, and anthropology. She worked with other young program managers excited to be at the museum, excited about science, and excited about working with children and enjoyed showing them that science was great and could be a career to pursue.

Working with Ellen Wahl Youth and Family programs Director at the Museum, they wrote and received a grant from the NSF program Innovative Technology Experiences for Students and Teachers (NSF ITEST) to expand their after-school program high school science program. Ellen continues to make exciting science programs to engage young students. Mandë also met her best friend, Dr. Jackie Faherty, at the museum. Jackie drove the Moveable Museum. Designed to go to underserved communities who found it difficult to travel to the museum, it is a travelling mini-museum fleet of buses with all kinds of things in it. There was one bus for dinosaurs, one for biodiversity, and one for astronomy. Jackie worked with the astronomy fleet and thought about applying to graduate school to become an astrophysicist, which she did. Mandë says Rockefeller taught her science, the museum taught her about education and public outreach. And both of them taught her the importance of and methods for serving our society by sharing science. Mandë thinks a good scientist should be able to share with the public what they learn about science in ways that the public can understand. The AMNH now has a PhD program in comparative systematics. It is a small program that takes about four students a year, and it is the only PhD-granting museum in the world.

During her year at the AMNH, Mandë was searching for a postdoctoral position. She wanted one that would bridge the gap between the work she was doing at Rockefeller and at the museum—that would be a bridge between chemical biology and the natural sciences. While a graduate student, she would walk across Central Park from Rockefeller to the museum and once attended a presentation by Dr. Baldomero M. Olivera (Toto) from the University of Utah, who showed a video of a snail-eating fish (these are venomous marine snails). Mandë thought the video was amazing and that it was crazy that a snail could eat a fish. She learned that these snails have venom similar to that of snakes—peptides in their venom can manipulate ion channels. So, she applied to work with Toto because she had found a project that would bridge her biomedical research with natural science. The snail's peptides have biomedical relevance because compounds in the venom can be used in drug discovery.

While she applied for the postdoctoral position in Utah, she also applied for a Science and Technology Policy (S&T) fellowship at the American Association for the Advancement of Science (AAAS), which was about applying scientific acumen in science and technology policy and international affairs. Mandë got the one-year fellowship to go to Washington, DC. She had come full circle, and, at last, could do international work, sharing her scientific expertise and learning how science can contribute to federal programs both nationally and internationally. There were many work placement options, and she wound up at the NSF, where she met Dr. Kerri-Ann Jones, a force in the international science policy arena, who became her mentor. Mandë became a science ambassador and went to Scandinavia, where countries spend 3.4 percent of their gross domestic product (GDP) on research and development. Then she went to Latin and South America where they spend about 1 percent of GDP on research and development. She found that government support for science and education helps lift up society in general, and countries can thrive. She learned that it is important for scientists to engage both globally with each other and with policy making. Policy makers decide on science funding and sometimes which areas of research will be permitted—for example, which stem cells lines can be worked on. Policy makers need accurate, unbiased, information. The world now faces many challenges that are borderless, ranging from infectious disease to food scarcity to energy. These are scientific issues and require that science and technology practitioners engage with policy makers to come up with solid, robust, sustainable responses. Her experiences with AAAS led Mandë, together with Jesse H. Ausubel, Director of the Program of Human Environment at Rockefeller, and Rod W. Nichols, Executive Vice President at Rockefeller, to develop a science diplomacy course that she now teaches. Held at Rockefeller, it is open to the City University of New York (CUNY) and AMNH graduate students After she developed the course it had to be approved by the Dean. When they pitched the course to Dean Sidney Strieckland he approved the course. Their science diplomacy course, which received funding from the Hurford

Foundation, is the first of its kind and was designed to encourage graduate students and postdoctoral scholars to think about science globally: for example, if planning experiments to develop a vaccine, think about how to solve the problem with a non-western cultural outlook and access to different equipment.

Students come from all over New York to learn how their science can be used on an international scale. Science diplomacy has three components: one in which policymakers enable science (like the International Space Station); a second in which scientists advise policymakers and give their expertise on important matters (like science advisors to the president or secretary of state); and last, confidence building between nations in conflict (like the Israeli-Palestinian Science Organization). Science has a culture of truth and integrity; scientists cannot fudge their data when they publish. Society has an inherent trust in scientists that is not always afforded to politicians. A major component of science diplomacy is keeping communication channels open so scientists can continue to work together even though their nations might be in conflict with each other. Mandë teaches that when science diplomacy is being applied in times of tension, it is important to keep science unbiased. No one wants the scientist to become a pawn for the government. She tries to expose her graduate students and postdocs early to thinking about science on a broader scale that can really make an impact. Science diplomats advise on a certain topic, share their technical expertise, and help solve problems. They shouldn't necessarily advocate for any special interest. If they do, it's imperative to make it clear they are speaking from a biased position, otherwise scientific integrity can be questioned. She thinks the more the public and policy makers know about basic research, the more support science will get and the less contentiously disputed science budgets will be.

During her AAAS fellowship, Mandë traveled with Harold Stolberg to Middle and South America. She went to the Republic of Panama and was visited the Smithsonian Tropical Research Institute (STRI) there, where scientists were looking for snails. There, Mandë decided what she wanted to do for her postdoc with Toto. She wanted to identify how the peptides in the marine snail's venom evolved over time to be so specific for their ion channel molecular target in cells. She received an independent fellowship for her postdoc from the NSF to work with Toto in his lab in Utah to learn about the biochemistry of snail venom. As part of her postdoc, Mandë went to the Paris Museum of Natural History to learn about the taxonomy and systematics of venomous marine snails. The expert there, Dr. Philippe

Bouchet, has an amazing malacology (mollusk) collection. Mandë also went to the STRI for her initial collection of snails. She found her first field work expedition hard, as she had no ecology training. Because most of the peptides that are found in snail venom are neuropeptides, she went to Berlin, Germany, to work with Ines Ibanez-Talon and learn about neuroscience. Neuropeptides act on neurons to manipulate cell signaling. The peptides in snail venom block neurons from transmitting signals to each other, such as the pain signal. In fact, the first drug on the market from snail venom is used to treat chronic pain in HIV and cancer patients. The peptides in snail venom block neurons from transmitting signals to each other, such as the pain signal. In fact, the first drug on the market from snail venom is used to treat chronic pain in HIV and cancer patients. It was discovered in the lab of Toto Olivera, Mandë's postdoc advisor, its commercial name is Prialt; the scientific name, ziconotide. This snail peptide, was found in the eighties Prialt the drug was approved in 2004. Made by Elan Corporation, it hits a new molecular target for a pain. When searching for pain therapies, most pharmaceutical companies were looking at opioids, which are addictive. This drug works on N-type calcium channels, so it avoids the opioid receptors completely, eliminating the issue of addiction. However, Prialt does not cross the blood-brain barrier, so it has to be injected via a spinal tap to get it into the central nervous system. Because the procedure is hard on the patient, Prialt is only used in extreme cases when morphine is no longer active or when the pain is very intense.

Mandë loved working in Toto's lab and going to his house for dinner. She thinks his wife, Lulu, is a great cook, second only to her mother. Toto was a great communicator and she loved his talks. He taught her to make talk about research a story and how to shape that story for different audiences—scientists, students, and the general public. Toto is from the Philippines and loves to give back. He collaborated with many scientists in the Philippines and has students from there come to his lab. He is a US National Academy member and a pioneer in the study of snail venom. Mandë had fun training with Toto during the two and a half years she was in Utah. They would have parties with wine at his home and sometimes on campus. While in Utah she learned fly fishing, skied, traveled to National parks, and made lots of friends.

At the end of her second year, Dr. Johnson called and asked if she would like to come back to do research and teach at CUNY. As she wanted to get back to New York, she said yes. When she first came back she was at York College, which is more of a teaching institution. Later she moved to Hunter College, which is much more research intensive. Although she thinks teaching is important, Mandë's primary interest is research. She is now a tenured Hunter College faculty member, where she studies evolution and learning from nature.

Prialt is so specific it targets only the N-type calcium channel and avoids things standard analgesic compounds do, such as being nonspecific and targeting everything. Mandë wanted to design smarter drugs with fewer side effects, similar to Prialt, but which were more applicable to the disease. She thought that Prialt's story illustrated how nature can help researchers find "smarter" drugs through the study of natural processes and wanted to find the clues in nature that would enable faster and more efficient result for cures. To do this, Mandë combined biology, chemistry, and evolution practices of the organ she was using to make the drug. One term for this interdisciplinary approach is "genomics," using, for example, the phylogenetic history of snails as a roadmap to highlight which individual species of snails are producing venom that may have biomedical applications. Venomics is a marriage between the millions of years of evolution of animal venom and new scientific technology, such as using next generation DNA and RNA sequencing and proteomics to learn about the genes and gene products expressed in the venom. If research can figure out how snails have so efficiently evolved peptides that specifically target ion channels, more effective drugs might be possible. This is the goal of Mandë's laboratory team; specifically, they are studying this in cone snails (predatory marine snails in the family Tererbridae) to determine how they evolve the peptides in their venom so effectively and how to use them to make smart therapies. It is interdisciplinary work, and Mandë has an appointment at both AMNH (working on biology) and Hunter College (working on chemistry), CUNY, and yes she's crossing the park again.

Mandë's team was the first to complete a molecular phylogeny of the terebrid snails—a family tree of how these snails are related to each other. This is used as a roadmap to identity which species of snails produce peptides that are bioactive or are of interest. For example, the Prialt drug is derived from the species *Conus magus*; so if you wanted to find a version of Prialt that crosses the blood brain barrier it helps you to know how *Conus magus* is related to other snails as more closely related species may have more similarity in their venom than less closely related species. This snail family tree it tells researchers that *Conus magnus* is related to, for example, *Conus geographus* by a certain distance, and they can find a branch on the tree that has all the similar cone snails to *Conus magus*. The hypothesis is that the venom on this branch will be similar across species. Mandë's team was the first to test this approach and use evolution as a roadmap to help decipher details on terebrid snail venom and direct research. Mandë's says of her research that it goes from mollusks to medicine or from beach to bedside. Her team goes on expeditions to collect snails, works in the lab to identify venom peptides, screens those peptides to determine their molecular targets, and characterizes how that target is being manipulated to control cell signaling, focusing on cell processes important to the human experience of pain and cancer

In the Holford Lab at Hunter College, the first peptide Mandë's students found and characterized completely was Tv1, from the snail *Terebra variegata*. Tv1 is very effective at inhibiting the proliferation of liver cancer; the patent for this peptide was the first for the lab. Her team is very excited about this peptide. It has been screened against cervical, prostate, neuroblastoma, and liver cancers, but it was effective only against liver cancer. It selectively kills cancer cells, not normal cells, unlike many current cancer therapies that can't discriminate between the two; when all the body's cells are attacked during cancer treatments, patients can become very sick.

Currently, Mandë's lab group is trying to determine the peptide's mechanism of action. Next steps will look at different stages of liver cancer to see if Tv1 affects cancer at the onset of proliferation, called metastasis, into other organs. Mandë is excited because by going from *mollusks to medicine* and tying evolutionary history to biomedical therapies, she is bringing together what she learned at Rockefeller and at AMNH. She also loves having her own lab and own group, setting her research vision, and writing grant applications to fund her work.

Her lab is split between three students of evolutionary biology and three doing peptide chemistry. Currently, two graduate students from the CUNY graduate program and a master's student from the NYU bioinformatics program do the evolutionary arm of the work, going on collection trips, building the phylogenies, and doing the transcriptome and genomic work.

Mandë has three graduate students doing the chemical biology work, and another New York University professor advises one student, Abba Leffler. One of the students works mostly on peptide synthesis, on methods to more effectively deliver peptide drugs. Because some don't cross the blood-brain barrier, they are exploring a Trojan horse strategy, putting the drugs in a nanocontainer and shuttling them across the blood brain barrier. Mandë's student has identified a peptide or a derivative of a peptide that seems to be selective for the alpha-3 beta-2 receptor in the brain.

Another student is creating computational virtual screening methods to process the huge database with information on the snail peptides that have been identified. (Each snail can produce anywhere from two hundred to a thousand peptides in its venom.) With collaborators from NYU and the team from Hunter the group is working on a manuscript. Mandë also has a postdoctoral scholar who has been with the lab since the very beginning and was instrumental in its development. She works both on peptide synthesis and drug delivery and is included in the patent for Tv1. Mandë hopes she will continue as a research associate when her postdoctoral work is complete.

Undergraduate students also work in Mandë's lab; working with amazing undergrads is one of the great things about being at CUNY. One of her students won first place in the Hunter undergraduate research symposium day, at which the undergrads from all the labs doing research present their work on a poster. Her lab had five posters, and the winning team was working on peptide synthesis.

Mandë met her significant other, Igmar Cuadrado, in high school where they were just friends. She was a tomboy, more into her classes and track team than dating. He was older, having followed the normal path through school. They used to commute together from Queens to Brooklyn, using the direct route to go to school but coming home by way of Manhattan so they could go to the Greenwich Village, hang out, and talk about sports and school. At the time the Village was a lot of fun. An industrial design major in high school, her boyfriend studied at the Parsons School of Design, received a degree from the California Institute of Arts, and is now an independent artist. They had stayed in contact and when he returned to New York, started dating. He would go to Rockefeller to attend the Friday lectures and then meet with the group post lecture and eat the cookies. Mandë and Ingmar are committed to each other, though not married, and have been together for almost seventeen years. Mandë thinks marriage is a contract but not a requirement, which may be either a cultural or generational thing, but that is not that important, because they are a family. Their first child, Ocean Luna Yumi, was born in July 2016; they tell her she controls the tides of her life as she commands the sea and the moon.

Mandë's elder sister is in education and started a school for special needs children. Her brother does computer engineering. Her younger sister is a lawyer in Washington, DC, and has worked in the mayor's office there.

When asked about being a woman of color in academia, Mandë pointed out that this is an exciting time to be one; the number of women in the field of chemical and physical sciences is growing. When she was at Rockefeller, Marnie Imhoff, Director of Development at Rockefeller, recognized the importance of trying to increase the numbers of women in science and started a Women in Science initiative, although there were few female scientists at the time, and Mandë was involved. Women scientists at Rockefeller had special funding and a special luncheon each year in May, celebrating women in science and the women at Rockefeller. Mandë now tells students who are interested in graduate school in the sciences not to worry about money because most schools pay them to go. She also tells her Hunter College undergraduates to apply to Rockefeller.

Mandë says that women are not the problem, the institutions are. They need to change in order to get more women in the field. They need to offer things like childcare so that women would not have to choose between a career or a family. Women should not be forced to compromise. Rockefeller has provided childcare for graduate students, postdocs, and faculty, so they don't have to spend extra time dropping off and picking up their children. She also likes maternity and paternity leave because men need to recognize the need for childcare and feel the same pressures as women. Her generation of male scientists want to engage with their kids from the very beginning, so they take paternity leave.

What does Mandë say to women interested in science? Engage and demand what you want. At Hunter, they are starting a Women in Science and Health (WISH) careers initiative, which will be similar to that at Rockefeller. It solicits donors and sponsors to make a dedicated development fund for supporting women in science and enabling things women want—institutions and organizations need to commit. Mandë received an NSF career award which she used to develop the Research Assisted Initiative for Science Empowerment for Women (RAISE-W) program to raise women up: early career faculty and women undergraduates work with executive coaches to determine the barriers preventing their success in STEM and find ways to get over those barriers and have a successful career, whether it be in academia, industry, or policy. Mandë suggests that women, whatever they want to do, should get a PhD and thrive in that area of study. RAISE-W is now in its fifth year at Hunter, with over a hundred students having participated, and she wants to keep it going. Mandë simply manages the program; the participants and coaches work directly together, confidentially. The whole group meets face-to-face three times a year to discuss common issues. The issues that women in science face are very similar to the issues that women in industry and the corporate world face. There are many: the imposter syndrome, or the feeling that I'm not good enough; how to get a promotion; how to manage time; how to write a paper or a grant; how to get tenure; how to negotiate tenure; and so on. Mandë thinks that if women learn how to tackle situations in the corporate setting, they can transition this to best practices in academic settings.

Mandë thinks that science is a good career choice that's open to everyone. There is so much to do and explore that we must be able to engage all of the citizens on this planet. She thinks we need to figure out how to use all of us to contribute to science in the best way possible. She encourages every young person, male and female, to develop their love of science, because she thinks it's innate. People are born with inquisitive minds. Children are born inquisitive, they take things apart, they taste things, they try to figure how to open things they shouldn't. She said not everyone has to become a scientist, because we need economists, we need politicians, we need doctors, and we need lawyers, but they need to be aware of the contribution of science to society. Everyone needs to think analytically about things, think critically about questions, especially about politics, especially about the funding of science programs. To advance science it is important that women and underrepresented groups engage because they can ask broader questions. It is very important for the enterprise that all be involved. To broaden the appeal of science, Mandë recently started a learning-games company, Killer Snails, LLC, with support from the NSF Innovation Corps (I-Corps) program. The first game, *Killer Snails: Assassins of the Sea*, won the award for Best Table Top game from the Games for Change organization. Mandë hopes that her company can celebrate "extreme" creatures, like venomous snails, by developing games and immersive experiences for students, teachers, and families in order to celebrate the amazing biodiversity on our planet.[7]

Notes

1. It was in 1960 that students of North Carolina A&T held the Woolworth Lunch Counter sit in, in Greensboro.
2. In 1969 the United States Department of Health, Education, and Welfare (HEW) began investigating the desegregation progress of ten states, including North Carolina, that had historically segregated systems of public higher education. HEW was responsible for enforcing Title VI of the Civil Rights Act of 1964 as it

pertained to educational institutions receiving federal funding. Title VI legislated that "no person in the United States shall, on the ground of race, color, or national origin, be excluded from participation in, be denied the benefits of, or be subjected to discrimination under any program receiving federal financial assistance." In 1970, HEW notified the University of North Carolina system that it was violating Title VI by maintaining a racially dual system of public higher education. In response, the university began drafting a desegregation plan. Eight years of draft submissions, revisions, rejections, self-studies, and negotiations followed. The matter was finally resolved through an administrative proceeding that began in 1980 and ended with a consent decree in 1981. Implementation of the provisions of the consent decree continued for a number of years, through 1988, during which time the university was required to make annual reports to the Federal District court.

3. Etta Gravely, interview by Jeannette Brown, November 13, 2012, Oral History Transcript, Science History Institute, Philadelphia, PA.

4. An instrument called a tachistoscope is used to identify where a flashed dot appears. The subject looks at a dot in the center of the instrument. The tester flashes a dot either to the right or left of that dot. The subject must say where it is.

5. Saundra Yancy McGuire, interview by Jeannette Brown, November 5 and 6, 2012, Oral History Transcript, Science History Institute, Philadelphia, PA.

6. Sharon Neal, interview by Jeannette Brown, May 2, 2016, Oral History Transcript, Science History Institute, Philadelphia, PA.

7. Mandë Holford, interview by Jeannette Brown, April 22, 2016, Oral History Transcript, Science History Institute, Philadelphia, PA.

4

Chemists Who Are Leaders in Academia or Organizations

4.1 Amanda Bryant-Friedrich

FIGURE 4.1. Amanda Bryant–Friedrich

Amanda Bryant-Friedrich (Fig. 4.1) is Dean of the College of Graduate Studies at the University of Toledo (Toledo).

Amanda was born in Enfield, NC, a small town about fifteen miles from the North Carolina-Virginia border. Her father was a farmer and her mother was a housewife. Her father only had a sixth-grade education and did not read or write much. Her mother graduated from high school in Enfield. Her maternal grandfather was a child of a slave and her mother was one of twenty-two children from two wives. They lived on a farm owned by a man named Whitaker. As her mother's family had been enslaved by the family that owned the farm, her last name was Whitaker. Amanda's paternal grandfather was a businessman who owned his own farm, on the other side of town. He was also involved in the illegal production of moonshine.

Amanda went to Unburden Elementary School in Enfield. Her first experience with school was dramatic, because she lived at the end of a dirt road and was really isolated from other families. The first day she went to kindergarten she saw all those little kids, and she was afraid because there were too many people there. But the daughter of her mother's best friend was there and invited her to come in to the classroom. Her first science class was in general science in fourth or fifth grade. She was so fascinated, she changed her mind about her future career of secretary or teacher and decided on science. Amanda went to Enfield Middle school in Halifax County, then the second poorest county in the state. The school had only basic infrastructure for science classes. She remembers her middle school chemistry teacher, Ms. Crowley, who told the students to put a mercury thermometer in a cork and Amanda accidently stuck it in her hand. They did not have much in the school, but her teacher taught her what she could. After middle school, Amanda went to Southeast Halifax High School, which was not a great high school but one of only two options available to her. She still was interested in chemistry and her high school chemistry teacher was Ms. Smith, a lady who always dressed like a lady. She would stand at the board and balance equations for hours. The students did not always understand what she was doing but they did learn a lot from her, as Amanda later discovered in college. Amanda was always doing something in high school. She participated in Hands Around Southeast which was similar to Hands Across America. This kept her out of trouble, she said. She graduated from high school as valedictorian.

Amanda's parents did not pay a lot of attention to her in school, because her mother was sick most of the time. Amanda spent much of her free time helping her aunt to take care of her mother and the house. She has one brother who is eighteen months older than she is; he went to college at Shaw University in Raleigh North Carolina.

During Amanda's junior year in high school, her mother started to recover from her long illness, and Amanda began to realize that she could go to college and started to apply to colleges, but without much conversation with her parents. She was accepted at Duke University and at North Carolina Central University (Central). Her high school guidance counselor told her not to go to Duke as she was not ready for that experience. So, she decided to go to Central, told her parents she was going to college, and packed up and went away because she had a full scholarship. Since Duke also gave her a full scholarship she considers that her decision to go to Central was pivotal and was the best thing she did. Central is an HBCU and at the time was experimenting with coed dorms, including Amanda's. The guys were on the bottom floor and the girls were on the top two floors. It was also an honors dorm. Amanda had a ball there because the students did a lot of socializing and studying together. Most joined a fraternity or sorority; Amanda joined the Alpha Kappa Alpha (AKA) and had "big sisters" living in the dorm who were also majoring in chemistry and biology and made sure she stayed on track. There were also Delta Sigma Theta sorority members there as well and she became friends with them.

Central was interested in preparing students as pre-med or pre-engineering students. Amanda was a pure chemistry major, so they had to tailor the courses for her. She took general chemistry and organic chemistry. When she took inorganic and advanced classes, she was the only student in the class. Some professors took that seriously and some didn't. So, when she entered graduate school, she lacked information about some aspects of chemistry because she hadn't had a formal class in the subject. She had to do a lot of learning on her own.

One thing that was extremely formative experience was internships with two major chemical companies: Dow Chemical and Merck Sharp & Dohme Research Laboratories. At Dow, she realized that she did not want to do large-scale chemistry because she did not want to tote heavy gallon containers of olefins. At Merck, she worked on angiotensin inhibitors and decided that medicinal chemistry was her calling. She also had great mentors at Merck, in Lansdale, PA.

Amanda also had a great mentor at Central, her organic professor, Dr. John A Myers, who invited her to work in his lab on a project supported by the NIH MBRS (Minority-Biomedical Research Support), which gave money to historically black institutions to train students from underrepresented minority groups. With that funding, Dr. Meyers was able to give Amanda, who was still an undergrad, a stipend, which she needed. She fell in love with the research—making heterocycles. As the labs hoods were not very efficient, bromine gas flowed in the air—they were brominating indole derivatives in order to add different groups. She was doing hardcore chemical synthesis even though she was only a sophomore in college.

Amanda had applied to Duke University for a PhD, because she had always wanted to go to there. Her mother was a seamstress and a woman who came to her home to have dresses made told Amanda she should go to college. This woman had gone to Duke and Amanda had never forgotten that. Therefore, when it was time for grad school she applied to Duke and to Ohio State University (OSU). She was admitted to both but decided on Duke for a PhD. Even though she liked the research at Central, she found out there were some holes in her education that she did not know existed until she got to Duke and studied with students who had a more rigorous coursework. It was a difficult experience; she completed the master's program, but then her research advisor was denied tenure. Amanda had to decide whether to stay and go into another group or leave with her master's degree.

Amanda's research for her master's degree was with Dr. Richard Polniaszek, now at Gilead Sciences Inc., making phosphorus heterocycles from dienes to make complicated aromatic systems. Her advisor was a thorough scientist who taught her excellent technique. Those compounds were used to make complicated aromatic systems. She wrote about this work in her master's degree thesis.

Amanda had made a lot of friends at Duke who helped her. One of them was Klaus Friedrich, who was in the United States doing a postdoc with Dr. Bertram Fraiser-Reid, Klaus helped her when she was struggling with coursework, having decided it was important that she do well. They became good friends and fell in love. When she had completed her master's and had to make a choice about her PhD. Since she had fallen in love with Klaus and he had moved back to Germany. Together they decided that in order for their lives to move forward as a couple and for her professionally, she should move to Germany to do her PhD, and she did. When she decided to move abroad she applied to graduate schools there. Because she had a master's degree from Duke she was at the same level as students going directly into PhD programs in Germany. Since her desire had always been to work on drug design to make medications, she applied to the College of Pharmacy at Universität Heidelberg in Heidelberg and also to a school in Mainz. She went to the Heidelberg school.

To learn German before she started her PhD work, she went to her boyfriend's family house and took German courses at the Goethe Institute. When she got to the research lab her colleagues wanted to speak English and she wanted to speak German. So, they compromised. She started in the department of pharmaceutical chemistry with Dr. Richard Neidlein as her research advisor. She did not do real pharmaceutical chemistry; she worked on his project, which was making complicated aromatic systems similar to her master's work. She was in Heidelberg for about three years, during which time she had to study other areas of pharmacy, pharmaceutical chemistry, technology, and biology. Her PhD thesis was written in German.

Amanda's most important mentors were her now husband and her father-in-law, who is also named Klaus Friedrich and was a professor of organic chemistry at the University of Freiburg. He would sit down with her when she came to visit and talk about the science she was doing and try to understand it. Navigation through the German higher education system was easy, because she did not have to worry about doing anything except her science. There was none of the other things that US graduate students have to do which might mean becoming a teaching assistant etc. She did not have to take formal coursework, because she had a master's degree, she just did an independent study. Her professors always spoke German, so she had to get up to speed in German quickly. Group meetings were also in German, so that when she went to her first group meeting she only picked up about 20 percent of what they were saying. But after a year she was able to understand everything and after two years, she could speak the language fluently. Since she was a perfectionist she did not want to speak unless she was saying things correctly.

Amanda decided she wanted to go into academia and teach, so, she needed to get a postdoc even if she and Klaus stayed in Germany. She called on her father-in-law to help her find the best place to go. Her father-in-law introduced her to Professor Bernd Giese in the department of chemistry at the University of Basel, in Switzerland, who was working with pharmaceutical companies, including CIBA, Bache Holding AG, and Novartis International AG, as part of an industry-academic partnership. Professor Giese accepted Amanda into his research team, asking her to use her phosphorous chemistry background to make modified nucleic acids. She worked there as a postdoc for two years.

Amanda married Klaus in 1994. By then, she had been living in Germany for a few years and her husband was working at BASF and his job had become permanent. Amanda was walking home one day and a bird pooped in the middle of her head—she had her hair in long braids down her back. She started crying and was still crying when she got home. Klaus cleaned the poop out of her hair and when it was clean, she stopped crying and he asked her to marry him. She was still working on her PhD at the time. They had an apartment in Ludwigshafen, which was close to the university and to BASF.

After she completed her postdoc, they decided to move back to the United States. Klaus got a job with BASF there, but Amanda had a problem because her degrees were from an HBCU, Duke, and a German university. She was told that she might have to do a United States postdoc, which she did not want to do. So, she took a teaching job at Wayne State University, first as a lecturer and then as part-time faculty member, teaching organic chemistry. She loved teaching, but Wayne State was not ready to hire her in a tenure track position. So, she applied for a position at Oakland University in Rochester, MI, teaching organic chemistry and the lab courses. There were two minority chemists in the department. Soon Amanda figured out that she was taking a big cultural leap because the staff members and support people were underrepresented minorities but none of the faculty were, except for one young man. The students wanted culturally competent staff, they soon found out Amanda was there and came to her looking for mentorship and help trying to understand how to get through the system. Amanda had not gone through the US system, so she had to figure out who her allies were and how to get information to give to the students. She became close friends with Christine S. Chow, is a biochemistry professor at Wayne State, with minority status, who knew a lot since she had gone through the US system. She trained Amanda on how to help those students and fight for them.

While she was at Oakland, the NSF was looking for ways to increase minority participation in science. Amanda, with a colleague, wrote a grant proposal to pilot a project to bring community college students into four-year institutions and to get them on track to get undergraduate and graduate degrees in chemistry. She subsequently became the director of the Center for Undergraduate Research, which brought in minority students from socioeconomically disadvantaged backgrounds who were at community colleges in the area and took them through the program. They began with a two-year grant and were later funded by the university, after the grant ended. Some of the students who went through the program went on to get graduate degrees, and others went to medical school. The NSF did not continue the funding because the number of students was small. But, since the program required a lot of hands-on mentoring and personal attention, it would be hard to work with a larger group.

Amanda was on the tenure track at Oakland and secured tenure on the normal timeline, despite being pregnant and bedridden and working from home. But the good news is that while she was home she wrote another NSF career proposal that was funded. She does say that women should stop the clock if they have the opportunity as that would be better for them. This means that they would not be expected to work for a period of time and then resume the tenure process when ready instead of having to work straight through the entire tenure time, even if you are sick.

Even though she was a tenured faculty member at Oakland University, with supportive colleagues and great students (mostly undergraduate and a few graduate students), Amanda decided to move to another university. She had been looking for grant money from the National Institute of Health (NIH), but didn't have the resources they were looking for and so needed to move to a more active research institution with more students and resources. She made the decision to move even though she had supportive colleagues and great students who were mostly undergraduates and a few graduate students. As a member of the NIH study section for cancer pathology, she met a researcher from the University of Toledo, who asked her to go there to meet some of the people and after that was recruited to move there.

When she moved there, she did not ask for tenure, thinking that she needed to prove to the world that she could get tenure at a research institution. She said that was a big mistake on her part. She went through the tenure process while pregnant again, with her second child. Her advice to young women is not to do that, but to negotiate tenure at a new job if they already have tenure at an institution.

At Toledo, Amanda's research group focuses on synthetic nucleotides and monomers with a variety of applications. They are studying the chemistry of cancer formation which comes from endogenous sources. They are looking at the mechanisms of oxidative damage to RNA that can contribute to the development of both Alzheimer's and Parkinson's disease. They also have a project that involves making heterocycles as neuroprotecting agents. At Toledo she was able to get a lot more resources and students than at Oakland. She says her research group looks like the

United Nations, for which she is grateful. She also continues to mentor undergraduates and she usually has about four undergraduates in her lab at a time. She works in the department of medicinal and biological chemistry in the college of pharmacy. Her dean Johnnie L Early II is African American and a mentor for her.

Amanda teaches medicinal chemistry to PharmD (Doctor of Pharmacy) students and also at the advanced level to chemistry graduate students. She also does a lot of outreach to young women, to help them if they are struggling with studies, etc. which she finds rewarding. Amanda is the immediate past president of the Association for Women in Science (AWIS) of Northwest Ohio. She has worked on several initiatives to try to bring more young women to STEM (science, technology, engineering and math) disciplines, is engaged in a Women in STEM Day which is held during the summer and has partnered with a local science center, the Imagination Station.

Amanda is currently the vice president of the Association for Black Faculty and Staff which she says is important to her because they are an oversite committee, and they pay attention to things that other people may not pay attention to because it does not affect them. They are concerned because the numbers of faculty of color have decreased significantly and want to find out why this is so. They have partnered with the university administration to make sure they understand that not having administrators of color affects both students and faculty. They work closely with the new president, Dr. Sharon Gaber, to try to improve this situation. They also work with the community, liaising with organizations like The Links—an organization of professional women of color, mostly African American dedicated to advancing women in education. The members of the Links talk to the university community to help them understand their position. There is a similar organization for men called The Boule is also a fraternity and became very interested in the University of Toledo because of low retention and graduation rates for African American males. In order to improve the retention rate these organizations have initiated programs like Brothers on the Rise for men and Talented Aspiring Women Leaders for the girls. Both organizations do a lot of mentoring of boys and girls trying to get STEM degrees.

There are a lot of pharmaceutical companies in Ohio in which students might think about working. Procter and Gamble Company is in Cincinnati, making personal care products. The University of Toledo has a degree program in cosmetic sciences that is very popular with students.

The university also works with small companies. The college wants to get the students in laboratories early to expose them to possibilities. Amanda is a liaison between her college and the Shimadzu Scientific Instruments Company, which gave the college a grant to buy cutting edge instrumentation now used to train students. She is passionate about giving students of color opportunities, because many people may not think they will be able to succeed. She works with all age groups. Because of her experience with internships while an undergrad, Amanda thinks it is important for student to participate in internships. Research Experiences for Undergraduates (REU) programs are one of the most important things that the government has offered to keep kids interested in STEM, she thinks. Offering students early opportunities to work in the labs that are doing cutting edge work will help keep them focused. She wishes the programs started earlier in students' freshman and sophomore years, while they are making career decisions.

During the American Chemical Society's Crystallography Year, Amanda's good friend Cora Lind went into an area high school to do the crystal growing competition. The students loved it and learned something new about chemistry that they had no idea about. The teachers still have students who are interested in growing crystals. The teacher in the high school got so engaged that she has continued the program every year with her students.

Amanda is also very active in the ACS. She is a member of the division of chemical toxicology and has served as program chair and councilor; as an ACS councilor, she represented the interests of her division at the governance meeting. She is also a member of the medicinal chemistry division, where she is on the long-range planning committee, which helps to design the programming for medicinal chemistry at the national ACS meetings.

Amanda is also involved with the Toledo local ACS section—through her partnership with AWIS because of the overlapping members and interests. The two groups run a lot of projects together in local schools and an Earth Day program at the zoo every year with experiments the kids can get involved in. The local ACS is also partnered with the American Association University Women (AAUW).

At Toledo, Amanda is the Director for International Graduate Student Retention and Recruitment for the College of Pharmacy. This has required her to travel to Saudi Arabia and Jordan a couple of times and also to Hungary to give talks and recruit students and form relationships. She has also spoken at a conference in France. She is a great relationship builder and also feels that she must talk about her work, because if you don't then it is just a hobby.

Amanda has authored three book chapters. One was about chemical toxicology. The other two were about women and minorities in STEM. She contributed the chapter "The Journey of an African American Female Scientist Scholar" in the book *From Oppression to Grace: Women of Color and Their Dilemmas in the Academy*, edited by Theodorea M. Berry, and about the struggles of women of color climbing the academic ladder, not just in STEM, but in many different disciplines. She also wrote a chapter for the book *Forward to Professorship: A Workshop*. This book came out of a faculty development workshop that brought women of color from STEM disciplines around the Midwest to Toledo to discuss isolation and how to help women who had chosen to go to institutions were there were very few people like them find a "tribe" to support them and be a community.

About two years ago, through her work with the AWIS, she co-wrote with Erin Cadwalader a chapter in the ACS symposium series on scholarship recognition from scientific societies. They reviewed data showing how many women were actually receiving awards in several organizations. The ACS was at the time starting to actively make some things happen and saw that as long as no one was really paying attention, it just all goes back to normal again. So, they have to be very careful about making sure that women are nominated for and receive awards. Even though Amanda was very busy, her children and her husband came first in her life. Though, she had decided early on that she would never marry anyone who did not respect her career. When she met Klaus, who came from a family of academics (his mother was a teacher and his father was a professor), she knew he would understand that it would be okay for his wife to have a career along with him. He was very supportive of her and urged her to get out and do things. They decided as a team that if they had children, they would become a part of their lives. Her children have had the advantage of traveling extensively and they meet great people. She has a son Klaus and a daughter Cornelia, who is nine years younger than her brother. Her daughter was born when Amanda was forty-one years old, so she was a late gift. Amanda has a lot of support, she hired someone to help her at home, because she could not do it all and keep her career going. Her kids have their Aunt Yvette to turn to when Amanda is not there. Her in-laws are supportive and they come in and help. Her mother also comes in to help. The key is to be intentional, she says. They don't make decisions in isolation, her career and kids have to be in unison, if they are not, she feels someone would not be happy. She helps her son with his homework and then she does her own work. She is able to manage work and family so that everyone is happy.

Amanda says her awards have mostly been for mentoring: she won the Alice Skeen's Award at Toledo for supporting women, and she won the ACS Stan Israel Award for Minorities in Science. But she also thinks her scientific contributions are important and wants to focus more energy and attention on those. To be a role model for the future generation of women and minorities and all people in STEM, she wants to make sure her science is top notch, and that means peer recognition. She says women and minorities spend so much time helping their community while they are doing science, that the science does not get recognized.

Amanda is also a member of the American Association for Cancer Research (AACR) and the Radiation Research Society to which she contributes a lot of science and training for young investigators. She is also a member of American Association for the Advancement of Science (AAAS).

Besides her scientific contributions to AACR, a lot of her work has to do with cancer etiology—understanding its causes. She was in charge of young chemists who were involved in AACR. The chemist in the context of cancer research is different because they are the ones who are creating new chemotherapeutic agents and understanding the mechanisms of cancer. Young chemists need a home in that organization, the Chemistry and Cancer Research focus group, and Amanda was responsible for the young chemists in that group. You have to make people aware that chemists develop new chemotherapeutic agents which are the things that are given to the patients. People need to understand their contribution to the field.

When asked about mentoring, Amanda says she finds it very fulfilling. There are many young people who deal with things that we can't even imagine. If a young person in her classroom is preforming in the high As and suddenly drops to a C, she needs to wonder why, reach out to them, and ask what happened. That might be transient mentoring, but it is just as important as long-term mentoring for students, who may be balancing two jobs or be single parents and may have no money but are trying to hold it all together. She even has to mentor students who come from the best families but don't know anything about being a chemist. It is not just about underrepresented minorities or socioeconomic disadvantage, it is about people who don't understand the chemistry discipline and are missing out on that opportunity.

When asked about the downsides of her career, Amanda said there was not a single step that was not hard. From the second-poorest county in North Carolina she arrived at Central with twenty dollars given to her by her parents. She studied with lots of other students who had more than she had, but they did not take education as seriously as she did. She said she struggled and fought and went to classes. She was able to find people to help, which made her undergraduate years good for her even though she had to work very hard.

When she went to Duke, she found out that she was one of only three minority students in her grad-school program, and the other two had been educated at highly active research institutions. She was always treated differently and was always trying to catch up. Again, she worked twice as hard as everybody else in order to make it through. Again, she had good friends to help her, but she always felt she was not as academically good as the other students were.

She had a heated conversation with her research advisor at Duke when she started dating Klaus. He thought her priorities were wrong and did not understand why she was developing a relationship instead of focusing on school. He did not understand that she was fighting just to make it through the program and that this relationship was an important part of dealing with this struggle. In this aspect, he was not a supportive mentor.

She felt liberated when she decided to walk out of the program at Duke with a master's degree, because in Germany she had the opportunity to function as an individual. The people she met in Germany had more issues with the fact that she was American than the fact that she was black. In Switzerland, she did have to deal with racism and bias, because she was always seen as a brown foreigner. They assumed she had come from Africa and wanted to live there and take away resources from the Swiss.

Coming back to the United States and working at Wayne State was the hardest part of her career journey. People told her that they did not think she belonged there and the only reason she was there was because she was black. Her degrees and work did not mean much to them. Even though she was teaching there, and her department chair Carl R Johnson gave her a lab so that she could do research, it was difficult for people to accept her. So, after a lot of battles with them, trying to get them to understand they needed to give her an opportunity, she decided to go to Oakland. She basically quit one day. Even though Wayne State is noted for having a lot of minority students, the attitude of the faculty does not represent that at all. When she left, Amanda vowed she would never be treated like that again. What kept her going is that she refused to give up and she knew that there were students who needed her.

When asked what is her most important contribution to the field, she says that her work making oligonucleotides, to find small molecules that form in cells, is the most critical thing her group has done. They can actually determine products that come from chronic stress that can contribute to cancer; they can take a blood or serum sample from a person who has experience a high level of oxidative stress. By having an idea of where the stress is coming from in the body, what the source is, they feel they may be able to find a way to prevent a cancer-causing event that is going to initiate mutations. They do that by trying to find biomarkers of it, so that they know when that incident is occurring; then they want to see if they can find ways to treat the mutations. The hard part is trying to determine what molecules are forming and the pathways of that formation, but they think they can do that. Next, they have to see what the downstream effects are of the formation of DNA adducts, or protein adducts, that come from those molecules, and what they can do about changing it. They want to see if this stress is causing signaling pathways to be activated that cause cells to multiply faster. Why are these small molecules involved in that process? Then they would be able to find ways to inhibit the cells from multiplying so quickly.

Amanda decided to become involved in the board of the Maumee Valley Country Day School where her son and daughter go school. It is a prestigious private school, for students aged three to eighteen, many of whom come from very affluent backgrounds. She says that her children are now growing up in a bubble. She feels that for the development of children, parents need to poke holes in those bubbles and have some impact on the school. Some of the things she does in the private school tend to end up in the public schools as well. For example introducing the students to other students who may not come from affluent families or be of the same race so that they learn to get along with each other.

Amanda also reviews a lot of grant applications. She has been on review panels and committees for NIH, NIEHS (National Institute of Environmental Health Sciences), and NSF. The NSF requires that applicants describe the broader impacts of their work, which can mean different things to different people. Amanda thinks that broader impacts mean giving opportunities to everyone who comes to the table. She wants to make sure that all young people have the opportunity to be educated. For example, the NIH is there to make sure that we stay at the cutting edge of health and treating disease. They have to train people to do that as well. This information should be in the grant proposal.

Amanda has worked with students from the Middle East. and thinks it is interesting to see how similar Middle Eastern families are to African American families. Women are very important in these families, because they feel that education is important. The dynamic between men and women is also very similar. She has worked with two brilliant young women who came from Jordan and became professors in their home country after leaving her group. They were also wives and mothers. She learned so much just watching them and how they maneuvered through their world. She helped by transferring her belief that they could be mothers, have a family, and still be a scientist. Amanda has visited her former students in Jordan and seen for herself that they have wonderful families and are doing great in their jobs. This helped her to see that there is something that is transferrable, that we can show; these women coming from different cultures from us can use this information to be successful. The women she taught were both Muslim, but they were not subservient to their husbands, though the idea of a man's profession and a woman's profession were different.

Amanda has also been to Saudi Arabia where the culture is different, and men and women are educated separately. There are campuses for women, that even respect the difference that they have in teaching men and women and the culture that they have. She thinks that women of Western cultures can help women in Saudi Arabia become leaders in their fields and that this is something that the Saudi Arabian government will want to do, but increasing unrest in the area is making it much harder. When the government can spend time thinking about education, she thinks women will have a much more important role than they do now.

When asked about single-gender education in the United States, she said that this is an individual choice, because you have to have the right mindset to function well in that environment. However, she thinks there is a lot of value in it, and she was saddened to see that so few single-gender university-level institutions remain in the United States. More and more are becoming coed simply because of the lack of funds. She thinks that the relatively small amount of money going into higher education here is almost criminal because if the nation does not educate our people we are going to be left behind.

When asked about two-year colleges, Amanda replied that they are very valuable, but not everybody needs to have a college degree. There are people in this country who need to be able to do the things they want to do without going to a four-year university. She wishes there were more apprenticeship programs, where people could go and learn skills that are still required so they can get a job. There is a place for people who just want to use their hands to do things.

Amanda started her career at Oakland University in the faculty of Arts and Sciences. There, she was a chemistry professor teaching traditional organic chemistry. Then she moved to Toledo to join the college of pharmacy as a traditional medicinal chemist. There are ten medicinal chemists in her group. She teaches pharmacy students and students getting graduate degrees in medical chemistry. The profession of pharmacy is moving in a direction that is not as chemically oriented as it used to be. Chemistry is being integrated into the pharmacy curriculum in such a way that students don't focus on it as much as they used to. Pharmacists are now moving much more into caregiving. They give vaccines; they do medication therapy management and can tell you which medications should not be taken together. Therefore, the basic science part is becoming integrated in a way that it is not as standalone as it used to be.

Amanda is participating in an academic leadership fellows program through the American Association of Colleges of Pharmacy to become an academic leader in the profession. She is taking a close look at what it is students will need to know as pharmacists. For example, in a hospital, a pharmacist makes the rounds with the physician to make sure that patients are getting the right dosages and the right medication. It is a very interesting and very lucrative career—the fifth highest paid career now. As well as hospitals, pharmacists work in retail and industry. They are involved in drug trials and clinical trials. Students now enter college to get a degree in pharmacy as a major and graduate with a PharmD—few pharmacists now have a bachelor's degree. It takes six years to get this degree in most universities, but at Ohio State University it is an eight-years degree. Amanda was the first person to educate a PharmD-PhD at Toledo. In the future, she thinks there will be more students seeking hyphenated degrees, such as the PharmD-PhD, PharmD-MBA, PharmD-master's in public health, and so on. This would allow pharmacists to be integrated more into people's lives.[1]

4.2 Gilda A. Barabino

FIGURE 4.2. Gilda A. Barabino

Dr. Barabino (Fig. 4.2) is Dean of the Grove School of Engineering at City College of New York in New York City. She is the first African American woman to hold this position at a Predominately White Institution (PWI) and one of only four deans in the country.

Gilda was born in Anchorage, AK, on May 28, 1956, the second of five children. Her mother is Margaret Agnes Barnes Barnum and her father Norman Edward Barnum III. Her parents were from humble beginnings, high school sweethearts from New Orleans, LA. Her mother went to Xavier Preparatory School and her father went to Booker T. Washington High School, both in New Orleans. Her father wanted to go to college but could not afford it and instead joined the military as a career that would allow him to support a family—this is why Gilda was born in

Anchorage. Their children came in the order of two girls, a boy, and then two more girls.

Gilda's parents were staunch believers in education as a way to move themselves, their family, and the black community forward. Her mother, a stay-at-home mom throughout most of her husband's military career, later went back to school to become a nurse's aide. After Gilda's father retired from the army, he moved the family to New Orleans, took a job in the Veteran's Administration, and went to school in the evenings. When she qualified, her mom worked at Charity Hospital in New Orleans. Gilda's dad demonstrated his conviction that education is important, as he pursued a bachelor's degree in business administration from Southern University of New Orleans while working full time to support his family. Her oldest sister studied chemistry as a pre-med student and became a physician. Her other siblings became professionals outside of science. They are a pharmacist, a lawyer, an accountant, and a human resources executive.

While the children were growing up, the family moved all the time. Gilda does not remember much about Alaska, as the family had to move when she was eighteen months old, because she was born with a tumor on her nose which needed surgery and radiation treatment. Her father was given a compassionate reassignment and Gilda's surgery was performed at Walter Reed Hospital in Washington, DC. After that, they moved to Dover, DE. In Dover, there was no kindergarten. But Gilda's older sister was in school, and Gilda wanted to do what her older sister was doing. She would look at her sister's books and created a classroom for her sister's dolls. She taught herself to read. Her mother got her a library card so she could read books they could not afford to buy. Gilda even wrote on the wall in chalk, because she was imitating a teacher.

The military family moved again, and in second grade Gilda went to three different schools. The family was in Florida when Gilda was in the third and fourth grade; they were in New Orleans when she was in the fifth and sixth grade; she finished sixth grade and attended seventh grade in New Mexico. Finally, when her dad was near retirement, they ended up in New Orleans.

Moving around did not save Gilda from a variety of racially motivated experiences. When Gilda was in the third grade in Florida, they did not live on the base, where the schools were integrated. They lived in the town, where there was a white only elementary school a few blocks from her house. She was bussed to the black school, out in the country on a dirt road, where the materials were substandard. After completing seventh grade in New Mexico, her father was deployed to North Dakota but the family moved to New Orleans, where he planned to retire. Gilda finished school in New Orleans at Sophie B. Wright Junior High School for eighth grade and Benjamin Franklin High School for ninth through twelfth grade. In order to be admitted to Benjamin Franklin she had to take an IQ test and a placement exam. It was a public school but run like a college prep school. Gilda found out when the test was going to be held from a white friend of hers and shared the date with every black person she knew. Her mom gave her and other students a ride to the test. The people taking the test were surprised that so many black students knew about the test, but Gilda was the only one who scored high enough to be admitted. She found that out by being called to the principal's office over a loud speaker. People must have thought she had done something wrong. The principal had the list in her hand and shoved it to her, saying disgustedly that she had passed. This was the first time that Benjamin Franklin High school admitted ninth grade students and only two from her school were admitted, she and a white male student. Gilda realized that the principal was upset because a black girl had passed and would represent Wright Junior High School at Benjamin Franklin High. That's when she realized that the people who should be her advocates and who should be happy for her may not be. When she got home her parents were excited that she had qualified for admittance. While she was in high school, in the summer of her junior and senior years, Gilda also took courses at Xavier University of Louisiana (Xavier). She graduated from high school in 1974.

Gilda continued her college career at Xavier because she had so many credits there, was familiar with it, had a full scholarship, and was encouraged by her parents to attend there. She had spent most of her academic career in white schools, where she was either the only black in her class or one of a few black students. She was tired of an environment where white teachers were, at the least, not encouraging, and at the worst just the opposite. She was tired of white students pretending she could not be as smart as they were simply because she was black. In fact, the student who told her about the exam to get into high school did not qualify and regretted she had shared the information about the exam. High school counselors at Benjamin Franklin never exposed the black students to college recruiters. Gilda was a National Achievement Scholar with the kinds of scores that would usually bring recruiters running, but nobody recruited her, except Xavier, because they already knew her as a high school student. She feels that many of the Ivy League schools and other colleges started to recruit a wave of African American students in the late 1960s and early 1970s, but when she graduated, those colleges seemed to be placing less emphasis on attracting this group. She feels if she had been in another school in a different city she might have been part of a group that was recruited.

Gilda had studied some science in high school, and more at Xavier, but she wasn't encouraged to become a scientist by her parents, who did not have an understanding of science. They said college was "a given" but her major was her choice. Most African American students were encouraged to study medicine if they demonstrated skills in math and science. Her older sister was pushed toward medicine, but although Gilda started out in pre-med she did not think that was what she wanted. The clinical work was not to her liking, but she did not know what other science careers were available. In high school, her chemistry teacher was adamant that girls could not become chemists and paid more attention to the boys in the class. Gilda decided to become a chemist because she had been told that she could not do it. So, she enrolled in freshman chemistry at Xavier as a high school student. Her professor, J. W. Carmichael, had set up a modular approach to teaching chemistry, making it more accessible for all students. The class sparked her interest in general chemistry. One class feature she found helpful was the ability to go at her own pace through the modules. She was able to finish them in half the time as other students and began to tutor her classmates. Intrigued by that, and still unclear about a career, she decided to major in chemistry, with minors in biology and math.

During her freshman year at Xavier, she applied for and was accepted to a summer program at Harvard Medical School in Cambridge, MA— a summer research program for students who might want to become researchers. This was her first time away from home; her first time in a residence hall; and her first time alone in a new place. Her memories of the time are not very good. She was placed in the radiology lab at Massachusetts General Hospital, in Boston, where a postdoc was in charge as the faculty member was away for the summer in Israel. Gilda was there with another woman who had a bachelor's in biology and was going to be a lab tech. The post doc was not pleased that he had been left in charge and did not help much. He gave them some protocols to look at, they did some chromatography and ran some gels, but they did not get a real sense of the project. From that experience she figured that science was not all it's cracked up to be. If the purpose of the program had been to interest her in graduate research, it had failed her. But she also realized that when she was a high school student, she had done the same work as a woman who already had her bachelor's degree. She decided that she needed to find a path that was more challenging and fulfilling and determined to get a terminal degree, a PhD, so she could be in charge. She also decided that she would visit other labs that the summer program students were in and offer her services to the lab that was most receptive to her. She did so much helping that when one Principal Investigator (PI) had an event at his house for the people in his lab for the summer, she was invited. Whether the experience was good or bad, she had to make the most of it, learn what she could, and figure out how to use it later.

When she got back to Xavier she saw it with new eyes—it was not as grand and education as Harvard. But she became self-directed and realized you get what you put into it. She looked up the science literature in the library, found articles on her own for her studies and undergraduate research, and was able to finish at Xavier in three years. She still did not know what she wanted to do for a career. Her family wanted her to go to med school like her sister. At the end of her junior year, needing three more credits to graduate, she decided to finish her undergraduate degree at the University of New Orleans that summer.

She applied and was admitted to the LSU Dental School in New Orleans to begin studies as dentist, without knowing how she would pay for. She applied for an Army Health Professions scholarship in order to pay for her education and then thought that was a mistake, because the LSU Dental School was very racist. There was another black woman in her class and the white professors would not communicate with them, thus letting them know they were not wanted there. The professors said that black students usually had to repeat their first year. Gilda was concerned that she had chosen the wrong profession. The matter was settled when she fainted during observation of a dental procedure at which the patient's gums were bleeding. What finally convinced her to leave was a call to the dean's office mid-year when she was told she would be expected to repeat her first year. She resigned immediately. The dean called her a quitter and revealed he had not expected her to succeed. She told him off, replied she was not quitting because she could not succeed, but because dentistry could not be the profession for her as she did not want to be part of such a racist profession.

Because she had an Army Health Professions Scholarship, she found out that she would become a commissioned officer even if she quit dental school. She was assigned to Fort Lewis, WA, to a medical combat unit. Although she was newly married and her husband was attending college at the University of New Orleans, there was nothing she could do about it. Her husband had lived his entire life in New Orleans, but he said he would go wherever she went. So, they both went to Fort Lewis, and her husband transferred to St. Martin's College in Olympia, WA. The terms of Gilda's scholarship required a minimum three years' service even though she had participated in the scholarship program for less than one year. She had to learn about the military while at the same time directing people under her command. It gave her time to think about her future career. She decided to go to graduate school to become a chemical engineer, because she wanted to do applied chemistry. She figured this out from her reading, since there was no one to direct her or advise her. She was still interested in medicine, but not as a clinician and thought chemical engineering was the closest she could get to medicine. At the time there were no degree programs in biomedical engineering. Although her preference was bio-chemistry, the literature was clear that chemical engineers were getting decent jobs that would help take care of a family. (Since that time chemical engineering has broadened to include biological, environmental, material, and other areas of chemistry.)

Gilda wrote to schools saying that she was a chemist interested in getting a PhD in chemical engineering and asking what the entry requirements would be. Four of the schools she wrote to, Carnegie Mellon University, Johns Hopkins University, Rice University, and the University of Washington (UW), replied saying they would accept a chemist. She chose Rice because it was closest to New Orleans. Looking for financial aid, she found the NSF graduate fellowship, applied for and got it. She had no idea what a big deal that was. She had simply described the kind of research she wanted to do, and made a statement, without the help of a mentor or advisor.

Gilda was the first African American admitted to the graduate program in chemical engineering at Rice. When she arrived, she discovered that they had not been expecting a person of color, much less an African American woman. Gilda sensed that her experience was going to be different when she reported to the office on the first day to find where the orientation meeting was. The woman she met did not realize she was the new graduate student. During her first week, faculty members pointed out to her that her background was different, because she had a chemistry background and not a chemical engineering degree. Gilda found this strange, as she knew the program had accepted white male students with chemistry backgrounds. She did not feel unique. One faculty member told her they had never had a black student in the graduate program before. She asked him why that was important and told him she hoped he and others would support her. Later she learned that she was getting paid less than the other graduate students. The other students had gotten a raise and she hadn't. When she enquired about that, they told her it was because she had a fellowship. Later she discovered that she had always been paid less than the others. She enquired about that and again the response was because her money came from a fellowship. She wondered why they would penalize a student who was bringing in money that saved them from using their own money. She made her point and the university both increased her stipend and gave her the money she had not received. This is how she learned that one has to speak up or things don't happen. She also had difficulty with the course exams. Other students were sharing old exams and studying together. No one wanted to work with her, not even to compare solutions. After she had been struggling alone for almost a year, a guy who was about to graduate came to her to ask how she was doing. She told him she was worried about the exam the next day. He told her it should be the same as always. She said she would not know because she had not seen an old exam. He went home and brought her a stack of exams to study. She had one night to prepare, but she received her best score on that exam. In grad school, you have to get an A or B to remain in the program and she had been getting Bs. After she got the old exams to study, she started getting As.

There were about thirty students in her class, and they all went out together but never invited Gilda. One of the faculty members who taught kinetics appeared to be an out-and-out racist. There were only two women in that class and when she went to the teacher's office to pick up her exam, he handed her another student's exam that had a low grade. He was surprised when she told him that he gave her the wrong exam. Another time she was asked to take a prospective student to lunch. The student was another African American woman. Gilda didn't share her negative experiences but did invite her to join her in the school. The student did not come to Rice and wrote a letter saying how bad it was. Gilda was called into the office to explain what she had done. She told them she had done nothing, but she was glad the woman had written the letter because it gave her the opportunity to point out how hostile the environment.

The one thing that kept Gilda going at Rice was her husband. When it was time for her qualifying exams, he told her she had only one job at home, to study. He told her to stick it out, because she had the intelligence to succeed. He also said told her she was a black woman in science and he did not think she would necessarily be treated fairly at any school. Gilda felt he always had her best interests at heart and was not afraid to tell her if she was doing something she should not be doing.

Gilda had always wanted to use chemical engineering to develop medical applications in a situation in which she could be in charge. She knew the medical application had to be something that would impact African Americans, which is how she got involved with sickle cell disease. As a chemical engineer she studied fluid mechanics, such as the flow of a liquid through a pipe. Studying abnormal blood flow in sickle cell disease was like examining flow through a pipe—the pipe is a blood vessel and the liquid is blood. Now engineering principles are regularly used for medical applications, but at the time this was not common. Gilda's advisor at Rice, Larry V. McIntire, was already looking at using engineering principles in the study of blood disorders involving white blood cells and platelets. At the time Gilda was getting involved, the NIH requested proposals for new approaches to sickle cell disease. They wanted investigators outside of the hematology field who were not already doing research in sickle cell disease. Gilda's advisor had applied for one of those grants to look at abnormal blood flow in sickle cell disease, specifically the abnormal adhesion of sickle red blood cells to the vessel walls. Gilda became the first student to work on that project, although she did not start from scratch but used methods from other studies, adapting them to increase the understanding of that abnormal adhesion. This area of study became her career; many years later she is still doing this research in her current job.

Her thesis was on rheological studies in sickle cell diseases, studying the flow and deformation of the red blood cells in sickle cell disease. Normally red blood cells look like little doughnuts. The hemoglobin inside is basically in an aqueous solution surrounded by a viscoelastic membrane. The normal red blood cells are very deformable, folding over themselves and squeezing into capillaries smaller in diameter than the red cell. In sickle cell disease, because of the abnormal hemoglobin that will polymerize in deoxygenated conditions, they get stiff, can't deform or pass through the capillaries. Not only are they stiff, they also adhere to each other and to the cells lining the blood vessels. There is blockage because of both the physical trapping and stiffness and a flow slow down.

When the cells slow down they adhere to walls and impede the passage of other red blood cells. Approaching a study like this involves examining fluid mechanical properties to understand disease, as well as investigating the molecular mechanisms that account for these physical phenomena. Gilda started by investigating which molecular mechanisms account for the adhesion. She used video microscopy to visualize cell interactions, using an in vitro system (outside of the body) to mimic what is happening in the body. Now the work is even more exciting because animal models are available—transgenic sickle mice with genetically manipulated hemoglobin that express human sickle hemoglobin—and researchers can conduct studies using the intact vessels in the animal models that are not possible to conduct on humans.

The genetic aspect of sickle cell disease is related to the hemoglobin genes, which can be normal or sickle genes. In the blood, to the normal hemoglobin gene is referred to as A and sickle gene as S. The trait is asymptomatic and for the most part, people do not experience the symptoms of the disease. If you have one parent who is AS and another parent who is AS they are both carriers. There is a 25 percent probability that a child would inherit two AA genes, a 50 percent probability of inheriting AS (sickle trait) and a 25 percent probability of inheriting SS and manifesting the disease. It is thought that the mutation may have evolved as a protection against malaria. Carriers are protected because the malaria parasite does not survive well in sickle blood. Sickle cell disease is predominant in African Americans. But there are other heritages that have sickle cell disease including some Caucasians. It can be tested in newborns, but in some cases, carrier status is determined when people a do the blood test to get their marriage license. Then they are counseled regarding the chance of having a child with the disease.

When Gilda came to defend her thesis, she had shown herself to be a courageous, fearless fighter and researcher. She knew her area up and down like no one else would have known it. The people on her committee and in her group saw the kinds of things she did to move her project forward and to help others. She showed her resourcefulness for example when she needed umbilical vein endothelial cells and could find no commercial source. Since Gilda had passed her qualifying exams and was in the process of conducting her research for her thesis Rice helped her. Rice collaborated with the Texas Medical Center and all the hospitals in the area, knocking on door after door and setting up relationships with OB/GYN practitioners who did delivered babies. Gilda went to delivery rooms and waited to pick up the umbilical cords. Eventually Rice and the hospitals set up an agreement. Now her students use commercial sources.

Gila got her PhD in 1986 and then had to decide what to do. Again, she had no mentoring. She wanted to go into academia eventually, but she had never worked in industry and decided to give it a chance. She did not realize that once you get into industry it is very hard to go back into academia. When she started applying to industry, few companies were hiring. The best offer came from Rohm and Haas Company which liked her chemistry background and her PhD in chemical engineering. The company focused on the synthesis of acrylics and acrylates and knew her research had nothing to do with that, but they felt that since she was a researcher, she could learn. Gilda started her work with them in Spring House, PA, doing a lot of synthesis of acrylics. Then she moved to the Bristol Plant, which specialized in research and used research process engineers to do the job. It was hardcore chemical engineering work, taking a process from the bench and scaling it up with pilot plants and full-scale plants. Gilda scaled up everything from coatings (to put paints on walls) to cementitious coatings (cement coatings that were fortified with acrylics) and waterproofing for leather.

She especially respected the reactor operators at the huge full-scale plants. She was one of the few people to introduce herself as the engineer on a project and then seek their advice.

The people at Rohm and Haas were discovering that Gilda did not fit industry. She wanted to understand the science on a very basic level and that was not how they worked. And she missed research with biomedical applications. During her last year with the company, her group asked her to do some research and present to the group on any new chemistry coming up or on a new product, so that people could get on board quickly. She enjoyed researching in the company's chemical library and putting together presentations for the group. Her colleagues told her she looked like a teacher and suggested she pursue that as a career.

When she realized that working in this industry was not her cup of tea, Gilda found there were two ways to get out of industry into academia. She could either remain three years or less in industry (after three years her lack of publications would hinder acceptance by academia) or she could make a name for herself in industry after an extended number of years, so academia would hire her for her expertise and experience. Gilda was by then in her third year at Rohm and Haas, so she started writing letters and looking at journal articles and journal ads, and so on, for a way back into academia. She went to meetings of both the American Institute of Chemical Engineers and the NOBCChE Philadelphia, asked everyone if they had any openings.

At one of the meetings she met people who were department chairs who had openings and she gave them her card. Some wrote back to her; she started interviewing and got offers from Howard University, Prairie View A&M University, and Northeastern University. She took the offer from Northeastern in Boston, MA, because she thought her husband would have employment opportunities there.

Northeastern had a very small Chemical Engineering department with seven faculty; she was to replace someone who was leaving. The first course she taught was Chemical Engineering 1421, covering chemical engineering, kinetics, and reactor design. Over time, she rose from assistant professor to associate professor and also served as vice provost for undergraduate education before becoming a full professor.

In general, Gilda feels that even at Northeastern she did not have mentors, people to guide her career, sponsor her, advocate for her, and generally help promote her career. She was always finding her own way and dealing with people who stole her ideas and others who put her down. So, she sought people who could be supportive, developing a network of African Americans in many disciplines. She began with people at Northeastern, like Ronald William Bailey, chair of African American Studies, who reached out to her and told her about forming the organization when she was hired. Another person was David Hall, African American dean of the law school. In addition, to her work as an associate professor of chemical engineering, she was asked to serve on many committees. She also worked hard in the community by joining the minority networks for the New England Board of Higher Education.

All this networking led to another life change. The position of vice provost for undergraduate education was open. When David Hall was asked who would be a good person for this job, Gilda's name kept coming up. Because the law school is very independent, he did not know her. But he called her in to meet him, asked her to consider becoming vice provost for undergraduate education and to apply. Gilda said she was a researcher, not an administrator, had little personal respect for administrators because of personal experiences, and wanted to be a full professor before she became an administrator. He suggested she should really think about it. In her small department of seven faculty, five were planning to retire. This would have essentially left her alone in that department with heavy teaching so that she would not have time to do research. If she went to the provost's office, she would learn about a new area and work for an excellent provost; she decided to try it.

Although Gilda was afraid that she might not be able to get her research done if she took on this role, she applied for and got the job. As vice provost for undergraduate education she was able to do new things, create new programs, and build on existing programs. She enhanced the honors program, transformed a teaching center led by a half-time person by appointing a full-time teaching center director and providing it a physical space, and started an undergraduate research program. She realized that even if she did not like administration, she was good at it. And she found she liked building programs, reaching more students, and having more impact. She learned about all the programs in the college and attended conferences about the first-year experience, teaching, and honors programs, learning to apply the processes she learned at those conferences. She also was happy because she had a supervisor who supported her and gave her free rein to do what she wanted. David Hall also served as a mentor to her—perhaps her first mentor other than her husband—and continues to mentor her about the broader aspects of administration and higher education.

Lenard Brown, an associate vice provost for academic opportunity, reported to her. From him she learned that you can also be mentored by your peers. They worked together to reach underserved students exploring the reasons why this group was not already being reached. Were they underserved because they did not have the resources or because they were from underrepresented groups? Lenard and Gilda worked hard to help the college become more inclusive and serve all students. This experience broadened her expertise and helps her now in her new career. She is not stuck in her discipline. She has crossed disciplines and can publish in social science as well as chemical engineering, biomedical engineering, or medicine.

In spite of the fact that she was a vice provost, Gilda never stopped being a researcher. She wanted to be known as a scholar, felt that a person is more in touch with the faculty when also doing research, and more in touch with the students.

Gilda's first sabbatical in 1996 was at the Massachusetts Institute of Technology MIT, working with Robert S. Langer researching in the new field of tissue engineering. For her second sabbatical at Georgia Institute of Technology, hosted by the institute of bioengineering and biosciences,

in 2003 she wanted to fortify that work—Georgia Tech had a tissue engineering center funded by the NSF. Around 2006, Georgia Tech had an
opening and she was approached about going there permanently. She was
attracted to Georgia Tech. She had gotten to know the staff and the type
of research they were doing, and she was attracted by the critical mass
of researchers in her area at the institution. She appreciated the school's
reputation as a hotbed for sickle cell research; and knew the patient population would be greater there. She started in 2007 as a full professor, with
tenure when she arrived.

While at Georgia Tech she also became associate chair for graduate
studies. She took on that responsibility because she wanted to use her administrative skills to help enhance their graduate program and increase
its diversity. When Georgia Tech decided to put a greater emphasis on
diversity, they appointed Gilda as the first vice provost for academic diversity with an emphasis on faculty diversity. Gilda took that role for about a
year and a half but came to see that it was taking her away from research
and that the institution needed a full-time diversity officer. She was still
associate chair for graduate studies and was filling two academic roles.
She went to the president of the university, recommending a full-time administrative appointment, not a faculty member, in the president's office.
The university recruited and hired a vice president for institute diversity.
Since Gilda had already created a strategic plan, the new hire was able to
begin by implementing her plan. She thinks she helped create an entire
new diversity office, though without receiving credit for it.

Gilda was at Georgia Tech and not actively seeking another position
when City College of New York (City College) had an opening. People
there knew her because of her national leadership; the biomedical engineering department knew her as an outgoing president for the Biomedical
Engineering Society who was an outspoken, visible minority. Department
staff suggested the search committee approach Gilda about the deanship
at City College. She was not interested and told them she did not want to
be a dean because it would be difficult to continue doing research. They
asked her to visit and take a look at the college. Because of the mission
of the institution and the type of students they draw, she decided to take
a look.

She saw people with a commitment to excellence, who lived the school
motto. She saw a track record and a history. She saw students with a hunger
for knowledge and the opportunity to use it. The faculty had opportunities
in other places but stayed because they bought into the mission of the

college. Gilda had reached a point in her career when she was interested in having some impact and really making a difference. She realized that as dean she would be the public face for the school of engineering and in charge of setting an example, a vision, and a direction as well as impacting allocation of resources. She saw an opportunity to impact education and began to envision a school of engineering with a difference. She saw an entity of higher education immersed in its community and giving back to the community. She wanted to attract students from Harlem, a mostly black neighborhood in New York City. And she knew a mere faculty member could not make that happen.

At the interview, she was asked whether someone who had never been a department chair could be a dean. Gilda thought that was a way to put her down. She suggested she should be hired on the basis of her accomplishments: her ability to develop people and programs; her ability to get things done. She pointed out that she had had other roles with budgetary responsibilities without becoming a chair. They agreed and hired her.

As Dean of Engineering at City College, Gilda is using some of the things she learned as a vice provost. For example, in an effort to give her students a holistic education, she started a project with the Dance Theatre of Harlem asking engineers to come up with a project that they thought would help extend the life of the dancers. The students developed an app associated with pressure sensing-socks that could be put in a ballet shoe to measure pressure values as the dancer moves. The app allows a dancer to also record a reference video. These videos can be used to compare practice movements, see when movements don't match, and help identify movements that might make dancers prone to ankle injuries—one the biggest problems for any dancer. Gilda created the connection with the Dance Theatre of Harlem and went there with the students to watch the dancers in rehearsal and meet with the dancers to talk about what they do. This enabled the engineers to learn about what they might be able to do to help. The Dance Theater even had a little mini class for the students and had them doing ballet moves.

Gilda has brought high school students into her lab, as well as undergraduates and graduate students to do research. She tries to impart to her students a drive and zest for learning, because she knows the more you learn, the more you find out how much you need to learn. She constantly urges her students to strive for excellence in the classroom, in the lab, and in their careers.

Gilda had married her high school sweetheart in 1977 while they were both in college. While she was moving back and forth between jobs and locations so was her husband. At one time, they were living in two different places, her husband in Miami for his job at Florida International University and she in Atlanta for her job, at Georgia Tech while they still owned a house in Cambridge, MA, where she had worked at Northeastern.

Gilda's son Jori was born while she was in grad school in July 1985; she had finished her research and was working on her thesis, but if she did have to go to the lab, she took her son with her. Sometimes she put him in daycare, because it was hard to write while he was with her. Her son was one year old when she was hired at Rohm and Haas and attended daycare because her husband also worked. Both Gilda and her husband were active parents, involved in extra schooling, visits, trips, and other educational programs. Jori did not watch TV. When Gilda started at Northeastern, Jori was four and was placed in the campus preschool. Since Gilda was a new faculty member, she was able to have lunch with him at preschool. Since she could plan her schedule, she only took field trips if they did not interfere with her work. She baked cookies for the kids, which amazed the other parents. Her son "volunteered" at Northeastern when he was in school, even at the preschool. He worked in the lab with his mother but did not want to become an engineer. While they were living in Cambridge, he went to private school. As an undergrad he went to Northeastern. They were happy that he picked Northeastern over George Washington University in DC, because he could be near home and go tuition free. Although they still had to pay room and board, they wanted him to live on campus to enhance his college experience. Jori majored in accounting, became a grant administrator, and is currently at Tufts University. He previously worked at MIT and Harvard as a grants administrator, where people told him they knew his mom.

Since Gilda was partly educated in the south, she knows that "separate but equal" education did not work. She wants to make sure that everyone has an opportunity to receive the education they aspire to and believes that people should hold institutions accountable for making this work. She did a fireside chat recently and was asked how white people can help. She said they should share their privilege and power. When they see things that are wrong they should speak up. It should not only be the black person who speaks up—too often they get labeled as whiners and are ignored. Social justice is everyone's responsibility.

Gilda is trying to empower black women in particular, but also all women of color. The data shows that African American women are the most disaffected group and the most underserved group. They are the group that is doing the worst by all the metrics, especially going up the career ladder. African Americans do not occupy enough leadership positions, and they are less likely to receive grants. Gilda has held workshops for the NSF, including Emerging Frontiers in Research Innovation and other divisions within the Engineering Directorate. NSF reached out to her, because she had a reputation for developing initiatives to help minority faculty professionally and to become socialized into the profession. The NSF recognized that a disproportionately low number of minorities are PIs and they needed her help. She wants to run workshops that address all the layers of discrimination, individual and institutional. Minority voices need to be at all levels in science and academia and industrial. When Gilda goes to workshops, she finds that she is the first African American female who is engineering dean at a predominantly white institution.

Gilda is really interested in increasing the number of women and minorities in science using the humanistic approach. She said it is not just developing a program to, for example, expose them to research. There have been a lot of programs with limited success. Such programs need a network of people who want to really work with people who may not look like them. They all need to have conversations together about their research, if that is what they are doing, not just generalizations. Participants need to see people of color as researchers too. People of color also need mentors and sponsors.

As dean, Gilda tells her department chairs to select a diverse slate of candidates. They should do the hard work to find suitable people. It's not enough to just say they can't find people of color because they are not in the pipeline. For example, the Meyerhof Scholars program at the University of Maryland Baltimore County program is successful. It is focused on minority scholarship and awareness in STEM disciplines. The program has served as a model for fostering scholarship in the African American community. But this program can't just be transferred to another institution, because there might be contextual factors that make it work at the University of Maryland that won't transfer to a different institution.

All students, but particularly minority students, need support and encouragement at all levels starting with K–12 and going all through college and grad school. Currently it seems that, too often, students are left to sink or swim. Some get pushed out the door. For those that become faculty members, some may not get tenure. Women and minorities disproportionately don't get tenure. Students at every level need to know that someone cares about them and can give them encouragement and help them. Providing support and creating a sense of belonging is what Gilda thinks everyone, including parents, should do.[2]

4.3 Leyte Winfield

FIGURE 4.3. Leyte Winfield

Leyte Winfield (Fig. 4.3) is an organic chemist and currently chair of the Department of Chemistry and Biochemistry and Interim Associate Provost for Research at Spelman College in Atlanta, GA.

Leyte was born on October 4, 1975, in Baton Rouge, LA. Her parents, Lena Winfield and Lionel Brooks, were both in high school when she was born. Her mother works at St. Joseph's Catholic High School as facilities manager and personal assistant for the nuns. In this capacity, she oversees a number of renovation projects and logistics for various events. Her mother attended Southern University Agricultural and Mechanical A&M College for two years. Her dad is a contractor managing crews that clean up damage due to fire, flood, or other natural disasters. He went to college for two years and has an associate's degree.

Leyte is the oldest and has three sisters, Dennell, Lakendra, and Lashay, and a brother Lionel who is deceased.

When she was growing up, life in Baton Rouge was very slow, especially compared to what life is like now. Although it is the capitol city of Louisiana, life in her part of town seemed very rural. Kids rode bikes, climbed trees, and picked pecans and figs. Leyte would get in trouble for going into the bushes to gather blackberries, because snakes lurked there.

The schools were integrated when she attended. She went to a private, all-black Christian school, First Christian Academy, for pre-school. She was bussed to the central area of Baton Rouge for elementary school. At Tanglewood Elementary School, she thrived and excelled as an honor student. For middle school, she went to a magnet school, McKinley Middle Magnet.

In elementary school, Leyte was exposed to science through Arbor Day 4H programs and mini-projects. In middle school, she took the typical science classes—biology, earth science, and computer technology. The biology class, however, stands out in her memory. She recalls an experiment which required her to blow into a raw cow's lung to study air capacity. The exercise was to show how human lungs expand when they breathe. But she did not want to breathe into a raw and bloody cow's lung and got into trouble for refusing to do so. She said she learned the hard way that understanding science sometimes requires you to do weird things.

Since Leyte was born when her parents were in high school which means they were poor, they wanted her to have a better life than they did. Her parents stressed that she could and should be successful. Her mom made sure that Leyte attended schools and participated in programs that gave her the best chance of being successful, which often meant that she was the only black child in a mostly white classroom.

In the 1980s, the Baton Rouge school board piloted different theme schools, including ones based on medicine, the arts, engineering, and law. At that time, Leyte thought she might want to be a doctor and felt the medicine program aligned with her future goals. So, she decided to attend Belaire High School where she enrolled in the medical magnet program.

Leyte remembers her high school chemistry teacher, Ms. Cain, as stern, focused, and knowledgeable person, who did a lot to show her that focus was necessary to master the depth of content important to understanding chemistry. She thinks her high school experience taught her the fundamental chemical concepts, so much so that she was bored in college.

She had a similar experience with high school calculus. Her teachers provided her with a firm grasp of both subjects.

Leyte does not recall that her high school teachers did anything in particular to encourage her to go to college. Even the high school guidance counselors only got involved when they knew the student was interested in college. Because she was in the medical magnet program in her high school, class size was small. On the other hand, her parents were the motivating force that encouraged her to pursue an undergraduate education.

Leyte feels she always knew that she wanted to go to college, to do something that was intellectual. Peer pressure worked on her. When someone did something that made them look smart, she wanted to do it too. For instance, when Leyte applied to the Upward Bound program, she was not selected, and she was devastated because she wanted to be with her peers. Subsequently, she was accepted to a program called Project Success. Housed at Southern University, it was similar to Upward Bound in that participants attended enrichment courses on Saturdays, and tutoring was available during the week for additional assistance. During the summer, they stayed on campus with the Upward Bound students, a group of college-bound, high school students. Ms. Kendrick was the PI of the program. Participating in Project Success further fueled and sustained her interest in attending college. She participated in that program from tenth to twelfth grade and received the support needed to ensure her entrance into college.

Students began to choose a college in tenth grade. Leyte decided that she wanted to have an African American experience, since she had gone to largely primarily white K–12 institutions. She wanted to go to a historically black college and university (HBCU) and to a college with a small student-teacher ratio, because she felt she needed a lot of attention. She applied to Spelman College, Xavier University, Dillard University, and Howard University. Howard was the oddball in terms of size because it was not small. She was accepted to all except Spelman, which waitlisted her. She had wanted to go the Spelman because it was portrayed in popular media (movies and television shows) as a place for black women who wanted to be something. She did not realize that even though she was waitlisted for Spelman, she could have appealed that decision or applied later as a transfer student. Howard sent her a lot of things letters that said she was accepted, but, being a country girl, Leyte decided it was too far away and decided not to go. So, the choice boiled down to Dillard or Xavier. Dillard gave her a bigger scholarship and was just one hour away from home. She enrolled in Dillard and found it to be an amazing experience.

At Dillard, Leyte got the personal attention she was looking for and started chemistry in her freshman year. Her original interest in chemistry came from a desire to develop and improve consumer products. She wanted to make creams and lotions that were better than the ones on the market. In high school, she recalls a seminar in which one of the speakers was a pharmaceutical chemist who described his research developing medical products. That resonated with Leyte, as it aligned with her passion to create consumer products. The seminar reaffirmed her interesting in pursuing a college major in chemistry.

The benefit of attending a liberal arts institution is that it afforded her the opportunity to take courses outside the area of her degree. Making the most of this environment, she took a children's literature course in which she had to write a children's story about her career path. She found that fun. Another course she took was music appreciation. Leyte participated in the Dillard University Glee Club in her freshman year and took a few voice lessons. While most of the time was spent studying, Leyte also made time for some recreation.

Leyte completed a traditional chemistry curriculum with general chemistry, then organic, analytical, biochemistry, inorganic, and physical chemistry. Her professors were very involved in the students' lives. They checked on students when they did not show up for class and scolded them when they lost focus. Her cohort of chemistry majors started large but dwindled to about six by Leyte's senior year. Although Dillard was coed, her chemistry cohort was all female when she graduated. Her professors were actively engaged in her development though graduate school and her early years as a professor. She is still in contact with two of the faculty members, Dr. Scofield (now at Wooster College) and Dr. Stanford.

Toward the end of her freshman year Leyte joined the Reserved Officer Training Corp (ROTC) and the United States Army Reserves. Due to her concurrent participation in both, she qualified for a simultaneous enrollment scholarship which was awarded in her sophomore year; her merit scholarship was very good but did not provide all the needed funds to attend Dillard. Leyte thinks that joining the military was the best thing she could have done because it gave her greater focus. In high school, she had existed on raw talent. She could go to class, listen, take notes then rewrite the notes, and later be successful on an exam without ever reading the text book. Before taking college level organic chemistry, Leyte had not felt the need to even purchase any textbook. Her professor assured her she needed one for organic chemistry and she realized he was right. So, the combination of her experience in organic chemistry and the military training gave her the focus she needed to continue to develop as a scientist. Leyte served in the Army Reserves for fifteen years, resigning her commission in 2009.

Leyte had entered college as a chemistry pre-med major, because she wanted to get a medical degree (MD) and then a PhD in pharmaceutical chemistry. Her high school medical magnet program allowed her to shadow doctors in hospitals in her sophomore year. In her senior year, this experience led to a job in a dermatology clinic. She finished the medical magnet program as a nursing assistant. She remembers watching surgeries in the hospital, fainting at the sight of blood, and thinking she could not do that. Other things in the hospital disturbed her, but she still registered as pre-med/chemistry major in college.

When she told her chemistry professors that she wanted to go to the Pond's Institute,[3] they suggested she become an organic chemist. She struggled at first in organic chemistry, but later fell in love with it. Focused on finding a program in organic or medicinal chemistry for grad school, she considered a medicinal chemistry program at Florida A&M University or an organic chemistry program at the University of New Orleans (UNO). Her professors strongly recommended UNO, since the campus was getting a brand-new chemistry building and would probably have the resources needed for her success. Leyte received a Chaucer Fellowship for her first year at UNO and a teaching assistantship which lasted for two years, followed by a research assistantship. During her teaching assistantship, she taught general, analytical, and organic chemistry labs. A professor in charge of each lab was responsible for grading lab reports and creating and grading the lab exams.

Attending UNO turned out to be the best choice for Leyte for another reason—during her first year there her son was born. In New Orleans, close to home, she had the support a young mother needed. However, as a result of the pregnancy, she did not take the placement exams very seriously, underestimating the role of the exams in ensuring she could immediately enter graduate work. As a result, she had to retake some undergraduate courses and completed her PhD in five years.

Despite her ultimate success, graduate school was a different type of learning environment, presenting challenges unique to and separate from parenting. The school provided a nurturing atmosphere where one could fail, recover, and still make progress toward a goal. Leyte's advisor, Dr. Mark Truddel, encouraged her to get back on track, reminding her that mistakes happen and encouraging her to keep going. Her advisor was very patient, helping her develop the skills that she needed to be a successful organic chemist.

She started her research in the second year of grad school. Before selecting a research group, Leyte had to talk to all the professors in her interest area, organic chemistry. She ultimately selected Dr. Truddel for her advisor and was delighted to join his group, despite the fact she was two months from giving birth and could not start research immediately. The research required working with cocaine and similar compounds, which further delayed her research until her second year, as both she and her advisor felt it would be best to wait until her son was older before exposing herself to that drug or other laboratory chemicals. One year later her research was again interrupted for military obligations. This also prevented her from completing her qualifying exams until year three of her graduate studies.

Leyte's dissertation describes the design and synthesis of molecules that have the potential to reverse the effects of mind-altering drugs. The work sought to identify a drug that would reduce dependency on cocaine but would not be addictive. Ultimately, the hope is to identify remedies that will reduce the desire for cocaine as Nicorette does for nicotine.

Her thesis was titled "The Synthesis and Biological Evaluation of GBR12909 Analogues" and was dedicated to her son. The compounds described in the thesis were tested in rat brain tissue by Sari Increaser, a pharmacologist at the University of Miami. When a molecule was successful it would be tested on live rats and rhesus monkeys. Leyte made a series of analogues based on the successful compound GPR12909. In most cocaine derivatives, such as Ritalin and lidocaine, the tropane ring system of cocaine is replaced by a monocyclic ring containing one or two nitrogens. However, GBR12909 contains a pyrazine central ring in place of the tropane. Leyte made a series of analogues related to this compound using traditional synthetic approaches. The analogs were purified using chromatographic methods and characterized by NMR (nuclear magnetic resonance), mass spectrometry, and x-ray crystallographic analysis. She learned the power of collaboration early on. Dr. Edward Stevens helped her through the determination of the single crystal structure. One of his graduate students, now a physical chemist and friend, Zakiya Moore, ran the x-ray experiments.

While Leyte's dissertation work is quite impressive in most circles, she found it necessary to describe her work in everyday terms to make it of interest to family and friends. She summarizes it as the creation of cocaine abuse therapies, rather than as synthesizing multi-substituted piperazine derivatives.

Few individuals, let alone African American females, go to graduate school to study chemistry. Leyte started graduate school with two other African American women, but she was the only one of this cohort to graduate. Before Leyte, only one African American women had been in the program, some fifteen to twenty years earlier. In Leyte's third year of graduate school, another African American female, Florentina Payton-Stewart, entered the program. Leyte first saw Florentina in a physical chemistry course and was so excited to see another African American woman that she followed her to her car. At that first meeting, Dr. Payton-Stewart (as she is now) thought Leyte was crazy, but the two women have become friends. Leyte wanted to help her be successful by telling her all the things she had been through. Subsequently, two more African American females entered the program: Dr. Zakiya Moore, who currently works for the Department of Homeland Security; and Dr. Chanel Fortier, currently a lead investigator at the US Department of Agriculture in New Orleans, LA. Leyte wanted to support them by telling them about the classes, who to do research with, and which research groups would support them. She overwhelmed them with information. When she realized that she had even more information to share, she started thinking about becoming a professor.

When Leyte received her PhD, she was torn between becoming a professor and working for the US Food and Drug Administration (FDA). She eventually wanted to become a chemistry consultant for the FDA or the US Army, since she was still in the service at that the time. But she needed to do more research before deciding and applied for a few postdoctoral positions; although she interviewed with the FDA, she thought the postdoc would give her time to figure out her future. She found a position at Florida A&M with Dr. Kinfe Redda in his medicinal chemistry laboratory where she researched drug design, focusing on COX-2 inhibitors. Since she was in the School of Pharmacy, Leyte was able to do drug testing in house, including enzyme assays and hands-on rat testing. She began training to do the rat testing and when it came time to learn how to handle the animals, there were rats everywhere. The trainer picked them up by their tails and allowed them to crawl free over the lab table. Leyte saw that and left, never to return. She chose to follow a traditional postdoc path and did not teach while there. Although the postdoc lasted less than a year, she was able to complete one publication. About six months into the study, she discovered Spelman College was looking for an assistant professor in organic chemistry. Since she thought the opportunity was not going to last long, she applied for the job, eager to engage with those who looked like her and to give back. "I want to inspire excellence in future Spelmanites interested in chemistry," she replied when the Provost asked, "Why Spelman?" Leyte successfully interviewed and was hired.

When Leyte started teaching, it was a rough transition. She both lectured and supervised labs at Spelman. Some men from Morehouse College attended her organic chemistry class. Leyte was about twenty-six or -seven at the time and looked even younger. Everyday seemed to bring a new challenge. When she was wrong, she was wrong and when she was right, she was wrong. The students did not go easy on her mistakes and made her work hard to earn their respect. She recalls spending hours preparing for lectures during the first semester she taught; so much so that she developed total memory recall of both the information in the text and the specific page of the book she was speaking from. Her students' expectations and her personal drive for excellence allowed her to flourish as a lecturer and an innovator in the development of learning resources for organic chemistry. Her reputation still appears to be mixed; some love her and some don't. All in all, she believes she has earned the respect of many students, and she has established a sustained mentoring relationship with more than twenty students who have gone on to receive or pursue doctorate degrees.

Leyte was one of four writers of the Advancing Spelman's Participation in Informatic Related Research and Education (ASPIRE) grant application; a NSF-funded project for HBCU undergraduate programs designed to increase student and faculty interest in informatics—defined as anything to do with information, its generation, use, storage, or analysis. Leyte's research and teaching involves informatics. She uses molecular-modeling software to design compounds and to better understand their biological activity. The software allows her to see how the compounds might interact with a biological target, like a protein or an enzyme. She also uses the program to learn how to change the shape of the molecule to improve biological outcomes and medicinal utility. She called that cheminformatics.

The ASPIRE grant leveraged what Spelman was already doing: informatics-based research and incorporating the informatics into courses. It allowed faculty to build on that and emphasize their success.

The success of the initial grant allowed the institution to receive the second round of funding, which is focused on course-based undergraduate research. In Spelman was on the second implementation period of the grant.

Leyte was the second program director of Women in Science and Engineering (WISE), at Spelman; the first was Dr. Cornelia Gillyard. This program was initially funded by NASA (National Aeronautics and Space Administration), it is now funded by ExxonMobil which has sustained its commitment to funding engineering student scholarship. Recipients are selected during their freshman year to receive a full tuition scholarship for their entire undergraduate education (both at Spelman and the subsequent engineering school). The students must make continuous progress toward an engineering degree, including transitioning to approved engineering schools, in order to maintain the award. For example, if a student moves to Georgia Institute for Technology, the money goes with them.

As a part of the Southern Association of Colleges and Schools (SACS) accreditation process, Spelman has established a Quality Enhancement Plan (QEP), "Spelman Going Global," encouraging all students to have global experiences before graduating. To facilitate this for science majors, the college established the Global Science, Technology, Engineering, and Mathematics (G-STEM) initiative, which allows faculty to establish partnerships and collaborations to support student training and research. Leyte chose France for her partnership and she initially targeted the University of Strasbourg. Ultimately, she was able to connect to France through LSU, which had established student programs in Grenoble, France which has the Grenoble Innovation for Advanced New Technologies (GIANT) program. With this partnership, students complete all-expenses paid summer internships, an eight to ten-week experience allowing students to complete a research project while also getting a global experience. Two for the price of one! Students don't have to speak French, since English is the official language used in the research labs. Although students who can't speak French may find social life outside the lab more difficult, Leyte says most seem able to make the best of it.

For Leyte, receiving tenure was a long, grueling seven-year process. Receiving tenure at Spelman means you are excellent in teaching, scholarship and service as defined by your colleagues. The institution uses student course evaluations to judge teaching effectiveness. In addition, the department chair can observe the teacher in the classroom to add more context to student evaluations. The college has since added formal requirements for regular classroom observations by a committee of peers; every three years for tenure-track teachers; every year for non-tenure-track. Although peer observations were not a formal requirement at the time, Leyte asked her colleagues to come and observe what she was doing so they would have that data when they were evaluating her.

Leyte served one year as vice chair of the Department of Chemistry and Biochemistry, to observe roles and responsibilities, before transitioning to chair of the department. The process was somewhat competitive. Any tenured colleague in the department could apply. Interested individuals must be elected by their colleagues and approved by the provost for the appointment. Once she became chair, she shared her vision and goals for her tenure in the position.

Since Spelman does not have summer school, Leyte had the opportunity to pursue other development opportunities during this time and became a scholar in residence at New York University (NYU). The first summer, she did research to learn a new technique. She wanted to be able to characterize the biological activity of several of her compounds and worked with a group that was able to analyze the DNA-binding properties of a compound. Through a partnership with the principal investigator of this group and her sponsor for the residency, she was able to return to the program for two additional summers.

Leyte tries to publish as often as she has the time and energy to do so. This is usually in the summertime when she is not teaching and has resulted in several publications during her time at Spelman.

Leyte is a member of Iota Sigma Pi (women's chemistry honor society) and she is working to initiate an Atlanta George professional chapter. Eventually, she would like to induct senior-year undergraduates to the chapter.

In 2013, Leyte received an NSF Opportunities for Underrepresented Scholars grant to the Chicago School of Clinical Psychology. This was a one-year accelerated postgraduate degree for individuals who wanted to pursue leadership positions in academia. The fellowship was funded to increase the number of women of color, especially African American women, are not well represented in these leadership positions.

At the time she interviewed for this program, Leyte was mentoring three undergraduate research students and advising four senior chemistry and biochemistry majors. She adapts her approach to mentoring and advising to the needs of the individual students. In some instances, the focus is entirely on succeeding at Spelman. In others, the focus can shift to obtaining entry into graduate or medical school. At other times, she helps students figure out how to balance their social and academic goals. For several, she has continued to serve as mentor throughout their graduate education. She believes this has helped the students stay the course and keep going.

Formally, at Spelman her responsibility is to ensure her students are on track with their studies. She is a sympathetic voice to help them work through challenges they may be experiencing, both academically and personally. She helps them understand and make career choices; guides them in deciding on the type of degree to pursue and what summer program to choose. She helps them to understand that not every environment is made for them as a minority or as an individual. In some places, they can really excel and it can become the right place for them. Other places might be competitive and provide an ideal pedigree, but students may not get the nurturing that they need to help them develop.

Leyte is a member of NOBCChE and she was most active in the Atlanta Professional Chapter. In the past, the chapter conducted an annual Super Science Saturday at local middle schools, attended by three hundred elementary and middle school students. It opened with a science presentation, after which the students would be divided into smaller groups that would circulate through classrooms to get hands-on experiences with science. Demonstrations would showcase different aspects of science. An environmental scientist might utilize a stationary bike to explain calorie burning and how calories fuel different devices; they might make ice cream. The event would close with the students coming back together for a keynote presentation. One year, a taxidermist from the Fern Bank Museum of Natural History brought her animal collection to explain the process of taxidermy. Other Atlanta NOBCChE Chapter outreach activities include exhibits at the Atlanta Science Festival and career panels for students of color.

Leyte is also a member of the ACS, which she joined in 1997 while at Dillard University. The chemistry club at Spelman is an affiliate of the ACS. Students members have the opportunity to participate in outreach and career development activities. The club also prepares the students to present at ACS meetings or do outreach activities at the national level. Leyte has presented at a national ACS meeting and served on panels for both regional and national meetings.

As a member of those organizations, Leyte has connected with collaborators, including chemical educators. Although she is still doing research in the lab, she is transitioning to chemical education. To further her research in this evolving area, she is connecting with individuals in areas for which she has no graduate school training, such as statistics and framework development. She finds collaborations increase her skillset. She wants to understand the impact of teaching strategies using tools other than student evaluations. Specifically, she wants to know which resources promote student's cognitive development and self-efficacy and how they do so. She is also working to encourage faculty in her department to understand all current pedagogical methods and how to incorporate them into their teaching.

Georgia State University has a grant from NSF called Collaborations for Workshops in Chemical Science (CWCS), which funds one-week courses on a variety of topics. Leyte attends most often the workshop on project-based learning in organic chemistry. When she attended in 2006, she started thinking about how to reach today's students. She wants to make her classes more engaging, to take them beyond the textbook, and to find a format where students not only master a concept but can also demonstrate that they know how to use the information to create new information. She has presented outcomes of this effort on ACS panels and the Biennial Conference on Chemical Education (BCCE).

Leyte has not forgotten her early interest in cosmetic chemistry and plans to develop a course in cosmetic science. For the summer bridge program called Women in Science, Technology, Engineering and Mathematics (WISTEM), she created a cosmetic formulation project for entering freshman. Students were able to develop lotions and utilize general chemistry techniques to characterize their properties; they tested lotion's the conductivity and acidity related this to the hypoallergenic nature of some lotions.

When asked about her contribution to the field of chemistry Leyte spoke about her involvement in the development of breast and prostate cancer treatments. It is a significant area, but she is just one of the many researchers in the field. As a scientist, she has found that once one aspect of the problem has been answered, another question arises. Researchers have yet to figure out how to stop the development of many cancers. But she is proud of the small contributions she has made to finding a solution.

When asked if the undergraduate students of today are like the students of the past, she said, "That is the hot topic of the day. "At faculty discussions, there is much talk about not getting the same caliber of students. Current students are less prepared, or not as hungry as students once were." Leyte thinks they have the same potential as previous students, but approach learning differently. She feels it is the faculty member's responsibility to pull out a student's potential.

Leyte knows the volume of work K–12 students have to complete. The problem she sees is that the amount of work and the expectations set in the K–12 system do not translate into something that makes us feel we are doing better as a society. She says the system skips over fundamentals to get more breadth, but students don't have the ability to establish their knowledge at a deeper level of understanding.

The number of African American women professors in academia is significantly below the number of Asians and, perhaps, even Hispanics. Although African Americans and Hispanics used to be at the same level of career progress, Hispanics are trending up; African Americans down. Leyte does not speculate on the reasons for this trend.

Proud of her family, Leyte thinks her son is "the best thing ever." She feels that without his birth her first year in graduate school, she would not have had the perseverance required for graduate school. She did not want to fail as a mother or a graduate student. He was always with her except for the semester she was away for military training. She struggled, sometimes unsuccessfully, trying to balance her life when she was an assistant professor pursuing tenure. She had an exhausting schedule. Her son would go to school and after-school care. Leyte would pick him up at 6:00 PM, get him settled, and put him in bed by 8:00 PM. Then she would go back to the computer to write or prepare for class for the next day. She would make it to bed by 2:00 AM and be up at 6:00 AM to start over again.

Claiming he hates science despite assurances from his teachers that he is good at it, Leyte's son planned to attend Full Sail University in Orlando, FL. Leyte is trying to spend as much time with him as she can before he leaves for college. She has thought about marriage and more children, but it hasn't happened.

What advice would she give to a young, minority person about to pursue a career in science? "The path might not be perfect but one must be both dedicated to the path selected and committed to doing the work that needs to be done. Be clear on your goals. That will drive you over whatever hurdles you find: failure in class, unsuccessful interviews, or not getting along with your group mates, long hours, etc. You must stay dedicated."

A potential workaholic, Leyte does a lot of things to maintain her energy level and keep her life balanced. Although she does not cook often, her son thinks he is a chef. On occasion, she and her son cook together. She loves to dance: West African dance, urban line dancing, and even takes classes at Alvin Ailey's American Dance Theater when she is in New York City. She has tried running and enjoys hiking.

Leyte's idol is Madam C. J. Walker. She believes Walker was a cosmetic scientist with a focus on the mechanical manipulation of hair texture and the development of elixirs. Leyte has created a module for the organic chemistry course that allows students to think about how to renature hair, how to change the pattern of hair, how to alter the chemistries of hair follicles to change the texture and the curl pattern of one's hair. Madame C. J. Walker discovered how to take curls out, modern women want to do the opposite—put them back in. Natural hair, coiled hair, is a major trend that will last.

Leyte says by the time she grew up in Baton Rouge, LA, there was no such thing as Jim Crow laws. But the scars were very visible. It was clear you didn't go into certain part of town unless you were invited; you didn't go to certain people's houses if you were not of an acceptable skin complexion. It shocked her when a college friend admitted that Leyte would not be invited into her friend's home due to her skin color. Historically in Louisiana, Creole society separated itself from the rest of the African American community. Unfortunately, Leyte notices that subdivisions within the race based on skin color and hair texture still exist in popular cultures beyond Louisiana. She sees it in the hip hop culture and in reality television.

Leyte learned early that due to the color of her skin, she had to be smart to get ahead. In high school, she tried out for the cheerleading team. Because the school was only ten percent black she didn't get on the team. She overheard the coach say, "We already have a black girl on the team." Her response to that was to run for Student Government Association (SGA) president. She won. So, if authorities did not want a black cheerleader, they could have a black student president. The world she lived in always reminded her that she was black. But her mom always told her that this would not limit where she could go in life. Leyte was very ambitious as a kid and learned to focused on the things that would give her the edge: attending the Project Success program, running for SGA president, being very outgoing, and being successful academically.

She would tell students today to follow their passion. Don't go into medical school because you think it's cute. That will wear you out unless you have a passion for medicine. Be open to the different ways your passion can work out. She had a student who was interested in both chemistry and business, acquired a bachelor's degree in chemistry, went to work for Procter & Gamble, then earned her MBA. She was a wonderful example of how to connect your passions. Beware, however, not to mistake some mold you think you should be in for passion.

To conclude, Leyte says her biggest challenge was work-life balance. It's not something she excelled in. She realizes she could have used some help along the journey as a tenured professor, department chair, active researcher and scholar, and parent. She wants to encourage others to seek the help they need. "You really can have a well-rounded life, she says. "Remember, life may not be excellent in each area at the same time. We have different waves in our life. Sometimes we are really good at the professional stuff, and we're just okay at home. And sometimes we excel at home, and we're just good at the professional stuff. Look at it as if it's just the wave of your life at that time.[4]

4.4 La Trease E. Garrison

FIGURE 4.4. La Trease E. Garrison

La Trease Garrison (Fig. 4.4) is the Executive Vice President of the Education Division and serves on the Executive Leadership Team of the American Chemical Society (ACS), the first African American woman to serve in this position.

La Trease was born on August 8, 1972, in Sutherland, VA, Dinwiddie County, about thirty miles south of Richmond. Her parents are Larry Evans who is from Apex, NC, and Ruby (Williams) Evans, a native of Sutherland. While her parents don't farm, when she was growing up La Trease worked on her grandparents' farm, in the tobacco field, garden, and so on. Her sister, La Tanja Davenport, is six years older, married, and the mother of two boys. La Trease's parents still live in one of the homes on the family property. Her grandfather, Teamack Williams, was one of the prominent African American men in the county and owned and farmed his own land. He and his wife Corrine had seven children.

La Trease's father began his career in the United States Army after he left high school in Apex, and after he left the military, went to work at Allied Chemical.

Her father and mother met at North Carolina A&T, then a teacher's college Greensboro, NC. Her mother, while a student, was a civil rights activist, participating in several rallies and marches that ended in jail, despite her parent's efforts to get her to go home. She met Larry, a part-time campus security guard, when she was walking across the campus late one night, and he offered to escort her back to her living quarters. After they married, they returned to the family home in Sutherland where her grandfather had set aside a plot of land for each of his seven children. Eventually, four of the children chose to build their homes on the land. So, La Trease lived next door to her uncles, aunts, and grandparents in what was called the Williams territory. Her mother worked as an administrative assistant at Virginia State College (Virginia State) her entire career, retiring in December 2015 after forty-seven years and eight months there. She had also enrolled in the college and received a bachelor's degree. Both La Trease's parents were very active in their church.

La Trease's early life was spent in a rural environment. All the food came from the farm—vegetables and eggs. But she would not eat the chicken eggs until the family assured her they were the same eggs as were in the store. She remembers the opening of McDonalds and when the family first got cable TV. On Saturdays, she and local youths helped her grandfather on the tobacco farm, and she helped her grandmother cook lunches of salad and spaghetti.

La Trease learned a lot living in a rural environment. Her dad made sure his daughters learned how to cut grass using the riding lawnmower. He also thought driving a pickup truck was necessary for survival. This was considered very unusual for a woman to do, but La Trease also learned to drive her first car, a red F150 Ford.

La Trease became interested in science during her elementary and junior high school years. There was one high school in the county, one middle school, and about six or eight elementary schools. She started kindergarten at a private school on Virginia State's campus, a laboratory school attended by the children of college employees, but only for one year, because the school closed. She liked that school, because she was on campus in more of a family environment, which was different from the public school. When she attended first grade at Northside Elementary School, La Trease was ahead of the public school students in reading and writing. All the schools in Virginia were integrated then. But she remembers getting into trouble because she was bored and talked a lot. She spent her time talking and trying to help the teachers instead of doing classwork, because she had already covered it in private school. She thinks of this as "college experience" early in her schooling.

In elementary school, her favorite subjects were reading and math. She didn't have many African American friends. Her fourth-grade teacher at Midway Elementary School was an excellent role model and mentor. She took students in as if they were her own children. When she thought the students had leadership abilities, she gave them opportunities, even in the fourth grade. She would invite certain students to stay after school to help grade papers and to learn more about what it meant to be a teacher and guide to others. La Trease and her best friend Cari Carper were chosen for this role. This pairing might have been considered unusual at the time in the south, because her best friend was Caucasian. They worked together for Mrs. Goode who would take them home if the work went after hours. They were able to mentor and tutor the other students helping them to strive and do better in class. La Trease loved both the experience and Mrs. Goode for her example and encouragement.

Her next school was Dinwiddie Junior High School in Dinwiddie, VA, which was forty minutes from her house. Phone calls to school mates were considered long-distance, which made it hard to maintain friendships.

There, she finally was with other African American students. La Trease was a cheerleader from second grade through the eleventh grade.

Her eighth-grade science teacher, Cheryl Brown, was special. She was married to Kenneth Brown whose parents were friends with her grandparents. They all farmed together. Class was a close-knit family atmosphere. That year the county decided to hold its first science fair. La Trease didn't know what a science fair was, though she knew about science labs because she had seen them at Virginia State. Mrs. Brown started talking up the science fair and urging the students to participate. After talking to her parents about it, La Trease decided to participate and asked her mother to find someone at Virginia State to help her. Because she had heard people talking about it, she wanted to do a project on acid rain and its impact on farming. She was connected with Dr. Cornelius Lewis who worked at Virginia State in the agriculture and horticulture department. La Trease began her project in winter, but Virginia State had its own farm and a greenhouse, and so she was able to grow crops in the greenhouse. She grew corn, barley, green beans, and tomatoes. Dr. Lewis taught her how to make different concentrations of acid rain to use on the plants. He also taught her to use different levels of fertilizer so that they could determine the impact of the fertilizer versus the acid rain and how it would affect crop growth. When she showed her project to Mrs. Brown, the teacher was blown away. She could not believe that La Trease was able to conduct such a project. Mrs. Brown told La Trease had to make a display board. Her dad helped with that and La Trease took pictures of the crops and wrote up what was happening with each of the plants at various stages of their growth. She collected all the data and information—Dr. Lewis taught her the importance of keeping a lab notebook. He also taught her how to interpret the results and about the different plants and how they grow. She put this all together and went to the science fair. She remembers dressing up and standing by her poster board. The judges were scientists and other people who came to the event. The judges were blown away too. La Trease became the first winner of the county's first science fair. She was thirteen at the time. Her parents said, "You won, and you're African American. You won the first one in this county. You made history!"

La Trease went from the county fair to Longwood University for the Virginia Junior Academy of Science Competition, where she received an honorable mention. The county authorities were very happy that their first entry at the state level had done well. This experience sparked her interest in science. Her display boards are still at her parents' home.

After that experience, La Trease enjoyed all of her science classes throughout junior high and high school. In her high school, she was able to take AP chemistry and AP physics—new to the school at that time. Her teacher was Mrs. Bealer who also taught algebra.

La Trease was also very interested in business. She was active in the Future Business Leaders of America (FBLA) and served as the parliamentarian[5] and so learned Robert's Rules of Order. As a stickler for protocol, rules, and going about things in the right way this was a natural passion for La Trease. In competitions with the parliamentarian team, she made it as far as the state level. She was class president in her junior and senior year of high school. When she graduated in 1990, La Trease gave the class graduation speech which was written, without sharing with her parents. Her theme was a text from scripture that says there is a season and time for everything.

> For everything there is a season, and a time for every [a]purpose under heaven: 2 a time to be born, and a time to die; a time to plant, and a time to pluck up that which is planted; 3 a time to kill, and a time to heal; a time to break down, and a time to build up; 4 a time to weep, and a time to laugh; a time to mourn, and a time to dance; 5 a time to cast away stones, and a time to gather stones together; a time to embrace, and a time to refrain from embracing; 6 a time to seek, and a time to lose; a time to keep, and a time to cast away; 7 a time to rend, and a time to sew; a time to keep silence, and a time to speak; 8 a time to love, and a time to hate; a time for war, and a time for peace. (Ecclesiastes 3:1–8)

She felt that most of the people in her class were going to stay in the locality, go to a local college or not go at all, and she wanted them to understand that there is a time you must explore the world and this was the time to do that. She got a standing ovation and her parents were pleased. When she finished high school, she worked in the new CVS Pharmacy that opened that summer. She was intrigued by the pharmacists there and wanted to join the profession. She applied to Virginia Commonwealth University, Hampton University, Rutgers University, and Howard University and was accepted by all.

La Trease did not apply to Virginia State, because she felt she had been there as a youth. She had gone to the parties and the football games and had taken some summer classes, attending all their summer programs related to science, engineering, and math. Although her sister had majored in accounting there and her father assured her she could get a full scholarship, the school didn't have a pharmacy major.

She chose Howard University, because she had gone to a homecoming there with a daughter of a friend of her mother (an AKA sorority sister [Alpha Kappa Alpha Sorority, Inc.]) and she liked both the school and the city of Washington. Having come from the country, she thought it was great to be in a city. She had never been on any kind of public transportation before!

Howard had a summer program for any student majoring in the sciences, which began the Monday after La Trease's high school graduation and was designed to acclimate new students to the campus. It also gave them an introduction to zoology, biology, chemistry, and calculus. It was a six- or nine-week program and she had a chance to meet incoming freshmen from all over the country. Through the Campus Pals program, she was able to get a mentor whom she met her mentor before all the other freshmen came in. She was also able to pick her roommate for the school year. The program also allowed La Trease to find her way around the campus and meet the professors for her freshman year. She planned to major in chemistry with a focus to later attend pharmacy school. She found out early that her foundation in chemistry, based on what she had learned in high school, was not strong. She sought help from graduate students and professors to make sure she could get through the curriculum. She saw the difference between students who had come from the northeast coast and California. Their chemistry foundation was much stronger than hers was. This was also true of math.

La Trease was on the campus for about two weeks when she met D'Vell Medley Garrison, now her husband. His parents, George C. Garrison Jr. and Patricia Medley Garrison, also met at Howard; and two of his brothers also met their wives there. It became a family legacy. D'Vell was majoring in economics. He came from a Christian home—his father was a church musician and his parents were deacon and deaconess in their home church.

La Trease was able to study with her boyfriend and they took some of the same electives. They both liked theater and music so they took introductory classes in theatre and music. She thinks electives give you balance in your education. Her theatre professor sent them to the Kennedy Center for the Performing Arts as a part of the course. She told them how to dress and sit in the theatre which was great for La Trease, since she had never been to a theatre before. La Trease also took a course on the black diaspora, because she had not had many classes related to black history. Being at Howard was an eye-opening experience for her, because she had never seen so many African Americans before. She was in culture shock. Her boyfriend would say she said things differently from the way he was accustomed to: for example, she would say "going to make groceries" instead of "going to the grocery store."

The summers between her freshman and sophomore year La Trease was at home and had trouble finding a good job. She worked for a chimney sweeping company as a telemarketer. This was hard because people did not want to have their chimneys cleaned when it was 90 degrees outside. Other jobs were just as bad, so she decided it was best not to go home during the summer. Thanks to her boyfriend, La Trease started work at the ACS as an intern the summer between her sophomore and junior year. D'Vell had a copy of her resume and was driving by the ACS office. He stopped and asked the security guard if there were any summer jobs. The security guard sent him to human resources where he handed in La Trease's resume. She was called the next day and was told that *Chemical & Engineering (C&E) News* magazine had a part-time position open and they wanted to interview her. She said she only wanted a summer internship because her parents did not want her to work during the school year; they wanted her to study. But, she did go in for the interview and met with several people. She was called back and was told they wanted her to start and they would teach her about chemistry publications. They knew that she was not a lab person but a people person. So, she told them she would work from May to the end of August. She started out writing obituaries and learned a lot about prominent chemists. She was excited, because she enjoyed writing and she enjoyed the people that she met. She learned that the ACS was the top-notch place for chemists to work and to join as a professional society. La Trease ended up staying on after the summer was over fitting her work hours around her class hours. ACS headquarters was close to Howard, so she was able to commute back and forth easily. Her work did not interfere with her studies or her ability to be on campus. But she did not tell her parents about the longer-term commitment for a while.

At Howard, La Trease worked in the lab with Dr. Yilma Gultneh. He was one of her favorite professors as well as her analytical chemistry professor. She learned a lot in the analytical lab. She did not like sitting in lectures, but she liked lab because it was hands on. She made friends with her lab mates, some were in her wedding. She became the cover model for an issue of *C&E News* because they needed someone to pump gas in the photo. Later when she went to a professor's lab; he had posted that issue on the notice board and told everyone that she was his student. She sent copies home to her parents and her boyfriend's parents.

When Madeleine Jacobs became editor of *C&E News*, La Trease started working directly with her and found another mentor. Under Madeleine's tutelage, La Trease learned more about the editing world, writing about science, and explaining the people behind the chemistry. Madeleine taught her writing and communications style. La Trease was able to read and edit some of the letters to the editor, which gave her a lot of insight into what people thought about the chemical industry, the ACS, and what the ACS was publishing.

While still an undergraduate, La Trease acquired firsthand knowledge of what was happening even though she did not attend ACS meetings. She saw all the meeting programs and helped edit the regional meeting programs. She saw what type of science was being presented. She had information about chemistry that most students would not have, because she saw the business side of the field. As graduation approached, she was thinking about what she wanted to do and where she wanted to work. Her boyfriend had graduated the year before and became a flight attendant (he had a passion for traveling the world). La Trease approached Madeleine Jacobs and asked about jobs with ACS. Madeleine said La Trease had been a valuable worker so she would see. ACS decided to hire La Trease as a program assistant, so when she graduated from Howard, she became a full-time employee there, working with *C&E News* for two years. She later worked in the education division, which rekindled her passion for helping students. There, she worked with Terri L. Nally in the Academic Programs Office (Student Affiliates Office). She worked with the two-year colleges, the chemical technology program approval service, and the college chemistry consultant service. She consulted with Madeleine about the career transition, as she has with every move she's made at ACS. Madeleine has always been able to give her solid advice, not based on her personal opinions but on what she thought would help La Trease progress professionally, and because of her status in the ACS she has given her the best opportunities within the organization. La Trease really appreciated her insight and willingness to share input for decision making.

When she started in the Education Division, La Trease was the youngest person in the department. She thought that would count against her. She prayed for help to make the right decisions and asked D'Vell, now her fiancée, for help. She wondered if some of the challenges she faced were age based or culturally based? Did they have to do with her ethnic background? What would be the reason for others to suggest that they were not sure she should could handle her projects? But everything worked out well. One coworker later commented, "Generally, whatever La Trease touches, it turns to gold."

La Trease credits God for her talents and abilities. She also had many people in her life who have been great supporters and have given her opportunities to demonstrate her skills.

La Trease graduated from Howard in 1995 and got engaged the same year. She was married March 15, 1997, in Richmond in an African-themed wedding at the Fifth Baptist church. This was not her home church. The two families were so large they rented a building that would hold everyone. They had the reception at the Marriott Hotel in Richmond and her husband surprised her with a trip to Kenya for their honeymoon. For both of them it was their first experience outside of the country.

La Trease was working at ACS and her husband D'Vell was a flight attendant for US Airways—as well as liking travel, he wanted to make some money—until 2001; he retired after 9/11. As a couple, they were able to travel the world before their child was born. If her husband was stuck on the west coast La Trease would fly out on Friday and back home on the Sunday night red eye to go to work Monday morning. She feels that people should follow their passions while young and not put school and work first.

Their first child, Testimony Faith Garrison, was born December 4, 2001. She is named Testimony because there were some challenges during the pregnancy, so they prayed about it, and she came out just fine. After Testimony was born she joined her parents on their travels. D'Vell's father passed before Testimony was born, but he did know that La Trease was pregnant.

La Trease started her studies for her MBA in 1998 at Strayer University in Washington, DC. The campus was near her home, and she took one class a semester if she could. If she had to travel for the ACS, she did not take a class. Later she switched to online classes, which were a hybrid, some online and some face to face, depending on her schedule. She was not in a rush, because she already had a degree. She just wanted to learn how to run a business. She thought the skills of budgeting, project planning, and marketing would be good skills to know for her work at the ACS. She also received a certificate in editing from the Editorial Institute, in Alexandria, VA, and took a technical writing course at Howard. Her second child, a son, D'Vell Medley Garrison Junior, was born in 2004; her third and final child Tehillah Grace Garrison was born in 2008. She was named Tehillah, which means "sing a new song, sing new praises to God," because she was diagnosed with spina bifida while in the womb, but does not have it. She is perfectly healthy.

The ACS was very supportive of La Trease during her pregnancies, when she had to be out on bedrest or needed to take maternity leave. She thinks it is a great organization for a woman. She would work part time, or work from home as she was coming back from maternity leave. This worked out well for her as a young mother with a growing family. She did not want to be a stay-at-home mom. At first Testimony was put in day care, but she kept getting upper respiratory infections. So, they decided to hire a nanny after their second child was born and kept her until their youngest child started kindergarten. Her husband is supportive of La Trease, her work and her traveling. They work out together who will be home with the kids while one of them is traveling. They sometimes ask the grandparents to stay with the kids when they both have to travel. La Trease lives in Woodbridge, VA, about thirty miles from her office, so it is a serious commute. She slugs—this is an informal system of hitchhiking. A group waits in the commuter lot until a driver pulls up and announces where they are going in DC. La Trease tries to get a ride with someone who is going as close to the ACS office as possible.

One of the programs that La Trease was able to organize for ACS was the Pan-American Conference for undergraduates, held in Puerto Rico in November 2010. Students came from all the countries in the Americas for the week-long conference. There were technical talks and social events. La Trease worked with Dr. Ingrid Montes from San Juan Puerto Rico who helped her organize the conference. Bacardi Ltd. sponsored many of the events. Many of the international students had not been exposed to ACS, so organizing a professional conference at that level was rewarding for both the students and La Trease.

La Trease and Ingrid established a life-long professional and personal relationship as a result of working together on the conference. There is a long-standing joke between the two of them that when one makes a move into a new role (as a for Ingrid) La Trease would think that when she moves up in the organization Ingrid is able to move with her, because La Trease is an employee and Ingrid is an elected member of the Board of Directors. When Ingrid received the ACS Volunteer Service award in 2012, she acknowledged La Trease in her award presentation. They both think that they mentor each other.

La Trease was able to watch the Ingrid's struggles on campus in Puerto Rico, in terms of being a woman and moving up the career ladder. It was very evident that she wasn't as respected as she should be by the men in the chemistry department. However, Ingrid was able to introduce La Trease to the president of the University of Puerto Rico. La Trease toured the campuses in Puerto Rico and talked to students about careers in chemistry, what they could do, and about what the ACS does for chemists.

Another memorable program was speaking at a special conference for students in Puebla, Mexico, in 1998, at the request of Lyle Hall, an ACS member. La Trease did not speak Spanish and the students in Mexico did not speak English. La Trease went to Berlitz and took a crash, three-month course in Spanish. She was able to pick up enough to deliver her entire workshop in Spanish, even though now she still does not speak Spanish. Lyle was impressed. D'Vell went with her because she had to take a bus from Mexico City to Puebla. Taking a bus in a foreign land can be challenging, but they made it to Puebla. They stayed in a monastery near the university where the conference was held, then moved afterward to a local hotel for a few days of vacation.

She and her husband have a company called Pledge Stones, LLC, an awards and recognition company that only uses stones, of marble and granite for all its products. D'Vell is the CEO and runs it full time.

La Trease is a certified facilitator for four of the ACS leadership development system courses, as well as the Extraordinary Leader course, which she runs a couple of times a year as time permits. She has been to Thailand on a NSF grant to facilitate an Extraordinary Leader course.

La Trease moved from the Education Division to the Membership and Scientific Advancement Division, managing the Local Section Activities office. There are local sections all over the country that are arranged geographically. When a member joins ACS they are assigned to a local section based on their zip code. La Trease thought she had outgrown the undergraduates she had been working with in the Education Division so now she could work with the professional chemists. She put on workshops to help local sections leaders be successful and to encourage the creation of programming that would appeal to diverse populations.

She worked with local sections for many years and then assumed the added responsibility of working with the Technical Divisions of the ACS. The divisions are designed around the various technical areas of chemistry. In this role, she got back to national meeting programming. Before they had an electronic system to organize abstracts, she would take home a box of abstracts to put them in order as to date and time and group of the event.

When asked why ACS has so few members who are from minority groups, she said, "ACS has not and does not create the family environment that you often find in minority communities." ACS is all about science and the work and not about an individual's culture, as the minority science organizations are. She believes that ACS should introduce groups or meetings that focus on the culture of the person. However, things are changing and more minorities are coming in as officers of ACS. La Trease thinks minorities are just beginning to learn more about ACS and what it offers. She also thinks ACS is being promoted better to minorities now. She tells students that they have to be willing to take the step and reach out to larger groups and not just stay at their home base.

When asked about balancing work and family she said that being a wife and mother are extremely important to her. Her kids participate in activities and she tries to attend sporting events, practices, and school functions. She will work evenings after they are in bed to get her work done. She thinks other women don't think they can be a professional or executive because they have kids, but to her it is really a question of what one is willing to sacrifice to make it happen, what one is willing to compromise on. Because of work trips, she does miss some things; but technology is helpful because she can see something on line. She does not really take lunch breaks or do a lot of socializing in the office. She works so that she can go home to be with the family.

She does volunteer at her church, Mount Pleasant Baptist Church in Herndon, VA. Her husband is an organist there and he got her started in the youth ministry—he plays the organ for the youth choir, and she directs the choir. She has also volunteered for the audio ministry—she records the announcements both to be played on Sundays and broadcast on the church's website.

Being involved in the community is important to La Trease. She speaks to youth groups about how to be successful and how to chart a path through life. She tells young people to consider their passions in whatever they do, because they will be successful in something they are passionate about.

As for what she thinks her contributions to chemistry and to ACS are, she says, "students." Many of the students that she worked with in the Undergraduate Programs Office are now professionals. Many of them have PhDs in chemistry and are doing very well. She sees them at ACS meetings and they are now local section officers or serve on national committees. They tell her about how she contributed to their growth and development and how she introduced them to ACS. Two of them are married now with kids, and she remembers talking to them about how they could do have a family and still have a career. She says she considers herself like their grandmother, even though she is not old enough.

Most of the former students were undergrads when she met them and now work for major chemical companies. They are now serving ACS as volunteers and she finds that very fulfilling. One student was the first student liaison to the Society Committee on Education (SOCED). It was her idea to bring a student liaison to that committee since they were programming for undergraduates and needed to hear what students wanted for national meeting programming. The first student liaison was Samuel Pazicni who was an undergraduate at Washington and Jefferson College in Pennsylvania. He is now a full-fledged member of ACS, has been on the Younger Chemist Committee (YCC), and is now on SOCED as a full member. La Trease found it rewarding to see him grow.

She thinks that while she is not in a traditional career, she is in a place that most people don't think about. Being able to introduce students to not just ACS but other organizations that they could work for really opens the door for the generation coming up to recognize that it's just not about the chemical industry. There are other places that need chemists and need scientists and need that analytical thinking in order to be successful.

When La Trease was working on her MBA, she wrote a research paper on succession planning. She learned the benefits of ensuring that you groom the next people to come and help lead whatever organization or profession you are in. Therefore, she tries to coach people who don't report to her at ACS. She asks them first if she could make a suggestion that would help them. She has found that some people value her for being honest with them and helping them to grow in the organization. She talks to people about how they present themselves to others. Perception dictates how other people feel about them; people might perceive certain behavior, even if unintended, in a negative way. So, people should present themselves in the way that they want to be received.

When asked what she thought about the future of chemistry, she thought it would continue to grow. It will no longer be a profession of old white men, because more women and minorities are beginning to enter it. One time when she was asked to speak at Florida International University, they only had her name. When she arrived one of the younger men came to her and told her that he had expected an old white lady!

She thinks the face of chemistry will continue to change as our secondary system, high schools and middle schools, get more experienced science teachers. Then they will be able to introduce more students to science, which will increase diversity.

La Trease is the director of the professional advancement team, which oversees the career services unit. She wants to make sure the team provides the services that the chemists of today want, rather than just what they got in the past. She's exploring ideas, including having a major retailer sponsor a makeover session for the chemists, even a fashion show. This would help chemists dress appropriately for the job they will be applying for. ACS is becoming a diversified group of scientists. As the profession as a whole changes, ACS must change with it.

When asked about diversifying the leadership of ACS, she said the problem with volunteers is time. People may not have enough time to do the things that have been done in the past. It will continue to be an issue and will only get worse, which is a challenge for the organization. She thinks that some of the things now done by volunteers could be done by a staff person, if ACS could hire a chemist to fulfill administration or management functions. The association does have many chemists, including those with advanced degrees, on staff. Hiring more might help with the increasing volunteer burnout that they are beginning to see. She feels that the core group of volunteers may not have time to volunteer because they are working on both their careers and families. She wonders what ACS can do to alleviate that. Local sections are geographically based—maybe there should be virtual chapters for people who can't attend a local section meeting but would still like to participate.

She hopes in the future to be able to explore some of those new ideas. She now has a dual role and oversees technology for members as well as professional education, career services, and awards for the organization.

So, she is looking to see how she will be able to continue to advise the society for the chemists of today and tomorrow.

When asked about getting more minorities nominated for and receiving national awards, she said she has become the liaison to the Committee on Grants and Awards. She is talking with the chair of that group about what can be done, and they may have to reinvent the whole process, including the criteria and promotional efforts. One challenge is that minorities are not nominated for awards. It's important to find out why this is so. Then, if they are nominated but did not win, is it because of the criteria? Is it the selection committee? Does the selection committee need to diversify? She will be working on this.

When asked about the future, La Trease reiterated that people should stick to their passion and figure out the things in life on which they are not willing to compromise. Then they can ensure they are on the path to success. It's important to make time for yourself. Make sure you do the things you enjoy. She knows that when she leaves ACS in the afternoon she is going home to a place she wants to be and will have a good time. For women, having a supportive husband is important to their profession. Having a supportive partner in life will help you climb the corporate ladder.

La Trease was recently been promoted to her role as Executive Vice President of the Education Division.[6]

Notes

1. Amanda Bryant-Friedrich, interview by Jeannette Brown, March 16, 2016, Oral History Transcript, Science History Institute, Philadelphia, PA.
2. Gilda Barabino, interview by Jeannette Brown, August 5, 2016, Oral History Transcript, Science History Institute, Philadelphia, PA.
3. The Ponds Institute was in Africa.
4. Leyte Winfield, interview by Jeannette Brown, September 22, Oral History Transcript, Science History Institute, Philadelphia, PA.
5. A parliamentarian is a person to makes sure meeting are conducted correctly according to Roberts Rule of Order.
6. La Trease Garrison, interview by Jeannette Brown, March 6, 2016, Oral History Transcript, Science History Institute, Philadelphia, PA.

5

Chemists Who Work for the National Labs or Other Federal Agencies

5.1 Patricia Carter Ives Sluby

FIGURE 5.1. Patricia Carter Ives Sluby

Dr. Patricia Carter Sluby (Fig. 5.1) is a primary patent examiner retired from the US Patent and Trademark Office and formerly a registered patent agent. She is also the author of three books about African American inventors and their patented inventions.

Patricia's father is William A. Carter Jr., and her mother is Thelma LaRoche Carter. Her father was the first black licensed master plumber in Richmond, VA, and his father also had the same distinction in Columbus, OH, years earlier. Her father was born in Philadelphia, PA, and attended college. Her grandfather went from Virginia to look for work in Canada and became a stonemason. Later he relocated back to the United States, where he soon married in Boston, MA, and several of his children were born there. Later, the family moved to Philadelphia where Patricia's father was born. Her mother, who attended Hampton Institute, taught school and later managed the office for Patricia's father's business. Patricia's mother was born and raised in Richmond, as were most of her maternal relatives. Patricia had three brothers. They were all born during segregation in Richmond, the former capital of the Confederacy.

Patricia was born on February 15, in Richmond. She attended kindergarten through eighth grade in segregated schools that were within walking distance of home. In school, they studied from hand-me-down books, but her black teachers were well trained and well informed. They had bachelor's degrees; some had master's or even PhD degrees. To go to high school, Patricia took a city bus across to the east side of town, to the newly built school for black students, which incorporated eighth grade through twelfth grade. Her teachers were excellent instructors who lived in her neighborhood and knew her parents quite well. The teachers looked out for the neighborhood kids and acted as surrogate parents outside the confines of the home. Teachers and principals were also great mentors, dedicated to their craft; they encouraged students to understand the world and function as responsible adults. Patricia excelled in science and math.

She started seriously learning science in high school. Early grade school gave her a basic introduction to botany, such as how to identify the leaves of various trees and some fauna and other flora. In high school, Patricia took chemistry and biology, performing small experiments, enough to pique her interest in science. Her chemistry teacher Mrs. M. J. Williams, an excellent instructor, was kind and patient. When she was a graduating senior, Patricia was voted the most inquisitive classmate, which puzzled her. Mrs. Williams urged her not to fret over it because she said, "How else can you learn if you don't ask questions?" Patricia loved chemistry. She also took algebra, calculus, physics, and biology in high school and graduated as valedictorian.

Attending segregated public schools was normal then. Her parents were firm disciplinarians but nurturing, loving, and encouraging; they were dedicated to seeing that all their children attend college. The Carters' living room held bookcases filled with novels, instructional texts, and history and science books. One, called *The Industrial History of the Negro Race in the United States*, was authored by Patricia's great-grandfather Colonel Giles Beecher Jackson, a former slave who read the law under his employer and became the first black attorney to pass the bar in Virginia.

Patricia's house was in a segregated neighborhood and she and her neighbors were confined to activities held in black enclaves. They did go downtown to the white business area to shop in the major department stores, but they could not attend white theaters or sit at the lunch counters to eat and could not ride in the front of the bus. When they went shopping for clothes in the department stores they could buy them but could not try them on at the store. Some stores would not even let them in the front door. However, black families generally were self-contained when it came to civic and social activities, holding musical recitals, readings, and business meetings in their living rooms. The personal library in her family's living room was a useful supplemental resource for the public library.

There was only one public library for blacks in Richmond, in Jackson Ward, the neighborhood named for Attorney Giles B. Jackson. The black students could go to that library to study and complete school assignments. In Patricia's high school, there was a special black history course taught by Dr. Joseph Ransom.

Since Patricia's father was a master plumber she was taught the plumbing trade along with her brothers, who also became licensed master plumbers in addition to their designated careers. One brother, William, was a lawyer, another, Jerome, was an army colonel, and the third, James, made plumbing his career. So, mechanics and machinery became commonplace to her as well as doing things with her brothers such as modeling airplanes, performing chemistry experiments, playing with model trains, and taking care of numerous pets. She also took music lessons, taught Sunday school, had kitchen and home chores, and was a Girl Scout. She lived next door to the YMCA (Young Men's Christian Association) and the USO (United Service Organization) where black soldiers from the military bases like Fort Lee and Fort Eustis went on weekends. All in all, she lived in a male-dominated environment.

When Patricia graduated from high school she was awarded several merit scholarships to colleges and universities, and she chose Virginia Union University, a HBCU, because it was local and she could commute to college from home. Her parents were already paying room and board for her two older brothers who went away to college. When she got to university she did not know what to major in. However, when she met the freshman advisor who was a professor of chemistry she said, spontaneously, she wanted to major in chemistry with a minor in math. Patricia started her freshman year with general chemistry, then followed with inorganic chemistry, chemical analysis, organic chemistry, and biochemistry, but did not do much research in chemistry as an undergrad. She was a member of the Beta Kappa Chi Scientific Honor Society and Alpha Kappa Mu Honor Society and she received the Garnett Ryland Prize for Excellence in Scholarship.

Her professors were very encouraging, especially her physics professor who wanted her to pursue a career in physics, but she preferred chemistry. Her advisors were Dr. Herman Gist, who she thought was great, and Mr. Robert Walker. She had these two chemistry professors and other physics and biology professors. She thought they all were excellent instructors. They made her work hard and dig down deep to understand the material. There were about six or seven students who were chemistry majors so they got together to form study groups. All of the subject majors were African American, save one Asian student. Some of them went on to get higher degrees.

Patricia was at university from 1956 to 1960, and graduated magna cum laude. She decided to go to graduate school because of a third chemistry professor who knew of openings at other schools and encouraged her to apply to them. In 1960, she participated in a summer grant program led by Dr. Gist at Norfolk State College. Patricia and four students from other Virginia colleges researched a project in organic chemistry.

After her June graduation from Virginia Union, Patricia married an army draftee who was immediately sent overseas, thus giving her the freedom to participate in the grant program at Norfolk State. That fall Patricia went to Fisk University for graduate work in organic chemistry upon the recommendation of Dr. Gist. She stayed at Fisk for a year but had to leave because her husband came back from the army the following year and wanted to finish his own college studies. So, Patricia applied for a federal government job and went to work as a professional chemist at the Radiocarbon Dating Laboratory of the US Geological Survey under the Department of Commerce. She had to relocate to Washington, DC, for the position, which she was able to get because of her home training on the bench saw with her father and also because of her skill in a specific area of chemistry. She started her federal career as GS 5 with a salary of $5,335. Her plumbing skills came in quite handy since she had to saw wood into pieces to begin the carbon-14 dating process—these samples were brought in by geologists. Several years later, her daughter was born and shortly thereafter Patricia enrolled in night graduate school at American University. It was very difficult. She was one of three employees in her laboratory, the only woman, and the only African American. Two of the employees processed the raw material and the third was the supervisor. Many times, Patricia ran the lab by herself, because her geologist supervisor, Meyer Rubin, often was on an excavation trip somewhere in the world. At the end of each year, the lab's findings were published in the journals *Radiocarbon* and *Geophysical Abstracts*. All three staff were listed as authors and for some articles, Patricia was named first author due to the graciousness of her supervisor. One day the lab was visited by the renowned TV-journalist David Brinkley, who came to observe the dating techniques and the processing of raw geologic samples to prepare the radiocarbon isotope. Patricia thought it was a great job. While working in that building, she was also able to meet celebrities like Edward Kennedy and Rafer Johnson.

Patricia worked at the carbon dating lab for nearly six years and was cautioned by a chemist friend about the limitations of laboratory advancement if she stayed in one lab for more than five years. She was encouraged to leave for a desk job with the promise of a higher salary. The lab was also dangerous— when converting the raw sample into the carbon isotope C14, severe chemical reactions were conducted in sealed tubes. So, Patricia was delighted when a colleague urged her to try for a position at the US Patent and Trademark Office (USPTO), which she had not known about. He recently had been hired there. Because of the new EEO (Equal Employment Opportunity) laws, employment doors in all the sectors of government had opened for African Americans. The USPTO began to hire black patent examiners with degrees in chemistry, physics, and engineering. Through her friend's guidance, she was hired, and at a higher grade than her lab position. Later, her career accelerated under the tutelage of a high-ranking patent examiner, who was a white chemist and who became a dear friend. Patricia was often the only woman in her chemical unit. The patent examiner works within the confines of the federal patent system and examines a patent application, once it's filed, following the laws, rules, and procedures set forth by the Constitution, to see whether it merits a grant of patent. Patricia was trained for the job at the USPTO in-house academy. A registered patent agent is a person who practices before the patent office, and has clients who are inventors that apply for a patent. Patent attorneys and agents work in the private sector, and either work for corporations or have their own law office for the benefit of inventors.

In order to process an application, patent examiners had to understand the invention. The USPTO has three major divisions—chemical, mechanical engineering, and electrical engineering. Patricia worked in the chemical division. All patented subject matter is broken down to classes and sub-classes and then into smaller segments. The examiner is assigned to the area in which he or she is skilled. Patricia was placed in the organic chemistry division to examine applications in the pharmaceutical art followed by the bleaching and dying of textiles, which specialized art she had to learn. She then was reassigned to the thermoplastic polymer art and lastly to laminates in the chemical engineering area.

There are several types of patents an inventor can receive: Utility Patents, for inventions that have a use, covers about ninety percent of issued patent grants; Plant Patents are issued for asexually reproduced plants like roses, chrysanthemums, and the like; and Design Patents protect the ornamental look of an object.

Later in her career, Patricia became an expert in the Patent Cooperation Treaty (PCT), an international treaty that allows applicants from foreign countries to file a patent application in the United States under certain circumstances.

The USPTO offers extracurricular activities beneficial to career advancement. Patricia became a member of the board of trustees of the Patent Office Society; she was the first female member of the USPTO Toastmasters Club, a USPTO recruiter, and one of the first women on the federal women's program committee. She has been chief judge at various science fairs and president of the Intellectual Property Law Association. She joined the Capitol Press Club because of her journalistic writing. She became a member of the ACS, the American Judicature Society, the Afro-American Historical and Genealogical Society, and was appointed a Lemelson Center Advisor of the Smithsonian Institution.

Patricia was a founding member of the USPTO EEO Committee, which met monthly and held informative programs. This was the time when the establishment pushed for equal employment opportunities for African Americans and, at the same time, for women.

Patricia has two daughters. Her older daughter Felicia Armenta Ives is a computer engineer and IT consultant and her younger daughter Julia Alana Ives holds an MD and a master's in public health. Her first husband became ill and could not work for a number of years, so Patricia became the sole breadwinner, juggling career and family life. She also did community work and joined in civic activities—including joining the board of the local NAACP (National Association for the Advancement of Colored People). While working at USPTO, she applied for the position of Fellow of the Department of Commerce Science and Technology Fellowship Program. This was a one-year paid sabbatical that would give her the freedom to do anything she wanted. It was a career-enhancing activity. Patricia elected to be a staff member on the Subcommittee on Domestic Monetary Policy, chaired by Representative Parren J. Mitchell, MD, and drafted H.R. (House Resolution) 6735, a bill to produce medallions to commemorate Maggie Lena Walker, the first woman president of a bank, introduced by Congressman Mitchell. Ms. Walker was a close friend of Patricia's great-grandfather Colonel Jackson. However, the bill she drafted for Mr. Mitchell was not passed in the House of Representatives because it lacked the required number of votes by the members of Congress. Patricia had a seat in the House of Representatives for the State of the Union address by President Jimmy Carter—the actress Elizabeth Taylor was also sitting in the gallery with her. She was told that only the wives and husbands and high-ranking staff were so privileged. After six months at the Domestic Monetary Policy Office, she became a special director at the National Park Service for the latter half of the sabbatical, critiquing papers presented on a South Carolina sea island.

Patricia had her share of tribulation at USPTO. Promotions were difficult to get in the early years because of biases, discrimination, and racist attitudes. Her professional environment was overwhelmingly white and male. Her first supervisor in the chemical division was mean-spirited and dictatorial. He had the power to manipulate a person's personnel record by placing negative reports therein that were harmful to a career. Unfortunately, he did that to most of his employees. She was the only black woman in the unit under his supervision, though he disliked female patent examiners regardless of race. Women had to prove themselves to him. He was forced to take training under the EEO rulings, but nothing improved. When she was carrying her second child, he told her that her work was unsatisfactory and that she should look for another position. She was traumatized and pushed to get reassigned to a different supervisor—surreptitiously, to prevent any more discord. That changed the direction of her professional life. Her new supervisor was friendly, understanding, and supportive. She was able to advance, and she subsequently applied for allied positions associated with patent examining. She retired from instructing international patent application filing procedures to new employees, in 1998, with a salary high grade.

Patricia became intrigued with the variety of inventors from many walks of life and from various ethnic groups submitting applications to USPTO. She was a volunteer presenter at USPTO during the grand 1990 celebration of the bicentennial of the US patent and copyright systems. She was asked to present a paper about women inventors for the bicentennial proceedings and event. Back in 1890, there had been no women associated with the examination of patents or its prosecution, though women had been filling for patent grants then. They were not part of the system as examiners or as attorneys. But one hundred years later, women became a significant part of the patent system.

When she was exiting her federal position, Patricia went through a process to become a registered patent agent and was automatically given the honor because she had more than five years' experience as a patent examiner. She decided to register because then she could work for a patent-law firm, have her own clients, and file their patent applications. Traversing the patent system is very complex and difficult for the average inventor, and one of her responsibilities was to search patent filings for her clients to see if the invention had previously been patented. She has now retired from that position.

Patricia was able to have a career, support her family, and participate in extracurricular activities. Her extended family and good friends were very supportive. Although a single parent, because she was divorced from her husband, she put her two daughters through college and her younger through medical school. Her networking with colleagues and associates, family and neighbors was extremely beneficial to the growth and development of her career.

Patricia became a certified genealogist because she wanted to know about her family's lineage. When her older daughter was born, she was the first great-grandchild of two living relatives, one paternal and one maternal. Patricia started her family history search at the National Archives, in the Virginia State library, and by asking questions of the living great grandparents. She compiled the information, which went back into the 1800s and learned that she could apply for certification as a genealogist, which she received.

On February 9, 2012, she was asked to be the speaker for the Black History Month Observance at her former agency, USPTO, and gave a presentation on the history of African American women inventors to the agency commissioner and employees. She was presented with a certificate and was pleased to be so honored.

While a patent examiner, Patricia became intrigued with the diversity of inventors who come from many walks of life and from various ethnic groups. She accumulated a sizeable collection of information on African American inventors, and on women inventors and other minority inventors as well. She began publishing articles on them in the USPTO journals and other, mainstream, publications becoming an expert on the subject. In 1987, she self-published her first book, *Creativity and Invention: The Genius of Afro-Americans and Women in the United States.* When she retired, she published two more books on African American inventors and patent holders: *The Inventive Spirit of African Americans: Patented Ingenuity,* published by Praeger in 2004 and *The Entrepreneurial Spirit of African American Inventors,* published in 2011, detailing the businesses and enterprises that African American patent holders established with their patented inventions. Each book contains a comprehensive list of African American inventors. These titles are the most all-inclusive works that have been published on African American inventors to date and are available online. The inventions described in the books cover all endeavors and the minority inventors come from all walks of life, some unable to read or to write but nonetheless extraordinarily creative.

Patricia is now moving in another direction and has curated exhibits about African American inventors, one at the Prince George County African American Museum and Culture Center in Prince George County, MD, and the other in the Museum for Black Innovation and Entrepreneurship in Washington, DC.

Patricia thinks her most important contribution to her field is the information she has gathered and presented on the creative talent of African American inventors. She was the technical advisor for the USPTO film *From Dreams to Reality: A Tribute to Minority Inventors* narrated by Ossie Davis and filmed in one day at the patent office. Patricia, the overall handyperson, was the script editor, recruiter for extras, and key grip. It was tiring but a fun project and won an award at the Film Guild Association dinner. She has been on TV and radio advocating creativity and encouraging participation in the field of invention to enhance and better the community.

Patricia has tutored many grade school children for the AKA, instructing and mentoring them in math and science and giving them a message of hope and encouragement, because, she says, anyone, regardless of background, can be an inventor and at any age.

Patricia divorced her first husband and later remarried. Her second husband has a law degree but worked in law enforcement with the Washington DC Metropolitan Police, advancing to became the first black helicopter pilot at that agency. He is the first certified black genealogist and one of the founders of the prestigious Afro-American Historical and Genealogical Society. Starting in Washington, DC, with seven historians, genealogists, and archivists, it is now a national organization with about forty chapters. Her husband is the expert on the cemeteries and burial places of Washington, DC. He has written over fifty books, including the definitive and most comprehensive work on DC burial sites, *Bury Me Deep*. They met at a genealogical meeting at the National Archives.

Patricia is a past member of NOBCChE, to which she presented a paper. She also presented papers for the National Technical Association (NTA), and for Association for the Study of African American Life and History (ASALH).

She is the past president of the National Intellectual Property Law Association (NIPLA), an organization founded in 1973 by African American patent attorneys, agents, and examiners because they were not accepted in the predominantly white American Patent Law Association. She has presented many papers and seminars for NIPLA and also served as secretary and vice president.

In May 2015 Patricia received an Honorary Doctor of Humane Letters (DHL) degree from her alma mater, Virginia Union University, in recognition of her distinguished career in the area of intellectual property and her outstanding research and publications on minority inventors and patent holders. Her sponsor, Dr. Marilyn Tyler Brown, a university board member, endorsed her nomination as did a significant number of colleagues and dear professional friends. Patricia also gave an address about African American inventors to the university's history department. She is a member of the alumni association, which sponsored her at one of their meetings where she gave a presentation on minority inventors. She has numerous other honors and awards.

When asked what she would say to a young person about being a chemist, Patricia responded with words of encouragement. The field of chemistry has expanded into many diverse branches. She is intrigued by pharmacy and forensic science and thought more people should go into that field. But she said there are many avenues that one can pursue in chemistry, including the patent field whether as examiner, agent, or attorney.

5.2 Dianne Gates Anderson

FIGURE 5.2. Dianne Gates Anderson

Dianne Gates Anderson (Fig. 5.2) was an Environmental Process Engineer who worked at Lawrence Livermore Labs in California and created ways to clean up dangerous waste and contaminated sites; she is now retired.

Dianne grew up in a military family, so she lived and went to school in a lot of places. After retiring from the US Air Force, her father Major Norman Gates Sr. worked as an electrical engineer for Public Services Company (PSO) in Tulsa, OK. Major Gates's college degree from Tuskegee

Institute was in electrical maintenance, because they did not offer electrical engineering when he was there. However, he worked professionally as, and had the title of, electrical engineer, while at PSO. Her mother Eddie Faye Gates was a teacher, community activist, and historian. Dianne has three brothers and one sister. She is the fourth child and the older daughter.

Dianne was born in June 1959 in Warrington, England, and had lived in New Hampshire, Mississippi, and North Dakota before her dad retired from the military, when she was nine, and the family moved to Oklahoma where her parents had grown up. She started pre-primary school in England, when she was four because she was a curious, active child. When her family moved back to the United States, they were to be stationed in Biloxi, MI. However, a hurricane caused a lot of damage there, so they could not move into their new home and so stayed in Oklahoma near her grandparents for an additional six months, where Dianne went to a rural two-room school in Summit, Oklahoma. When their house was ready they moved to Mississippi, lived there for about a year, and then went to North Dakota for two years. Thereafter, she went to school in Tulsa, OK, from the fourth grade through high school.

She thinks she always liked science. She had some science in her elementary school classes. The chemistry class where things clicked for her was in the ninth grade at Edison Senior High School in Tulsa. She realized she loved chemistry better than math, and she took chemistry courses throughout high school. At her request, in her senior year the school created a class for her. She was a teacher's assistant in the chemistry classroom.

Her parents, and everybody else, encouraged Dianne to go to college; she just had to figure out what to major in. As a senior in high school, she participated in a "shadow an engineer for a day" program. But she did not get much of a sense of what chemical engineers do, because she was assigned to shadow a civil engineer. She had mistakenly marked the application form that she wanted to shadow a "C. Engineer," thinking that was the proper abbreviation for chemical engineer. As a National Merit finalist in high school, she was sent information and application forms by many colleges from all over the country. She decided to attend Oklahoma State University (OSU) because her brother was going there, and her family wanted her to stay in Oklahoma. She chose to major in engineering because she was good in math and science, and because engineers solve problems. She decided to major in chemical engineering because she liked chemistry. Also, she was told both by other students and by her advisor that chemical engineering was the hardest engineering major at OSU—she was interested in the challenge. At OSU, Dianne attended lectures and lab classes, the typical undergraduate curriculum. Since she is an early bird, she chose early morning classes so she could get her homework done in the afternoon and before the weekends, because she wanted that time to party and play. She did not join a sorority, because she thought it would negatively impact her grades. As an undergraduate, she was a member of a program called the Council of Partners (COP), designed to increase the number of minority engineering students. This opened the door to undergraduate summer internships and to a counselor, Danielle Mohanty, who worked with the undergraduate students in the program. Danielle proved particularly helpful when Dianne returned in her junior year after the birth of her son. One of the faculty members, Dr. Ruth Rebar, was also helpful by encouraging her. Dianne remembers going to Dr. Rebar's office one day and complaining that other students did not listen to her in class when she spoke or seek her out when trying to understand homework assignments, even when she often knew the answers. The professor said it had to do with how Dianne, an introvert, presented herself and gave her a book about body language and communication. This was the first time that she was made aware of how to groom herself to be look professional. She remembers Dr. Rebar's help with gratitude.

In 1982, Dianne graduated with a bachelor's degree in chemical engineering, after four years and a semester. She would have graduated in four years had she not taken one semester off (for the birth of her son in November 1980, returning to school with her six-week-old son in January 1981). After graduation, Dianne married her son's father and moved to El Paso, TX, where her husband was stationed at Fort Bliss Army Base. She looked for a job as a chemical engineer but could not find one. She believes her job opportunities were severely limited because she was a military spouse. In 1983 she had another child, a daughter. In 1984, she divorced her first husband and moved back to Oklahoma. After a long search, she eventually found a job at Amoco Oil Company, but realized that with only a bachelor's degree, her possibility of advancement there was very limited. At Amoco, PIs and senior staff with PhDs had private offices on one side of the hallway. Other staff worked at desks in the labs on the other side of the hall. Dianne felt that she could not cross the hallway without an advanced degree, and she realized that there was much more she wanted to do besides handle the laboratory part of someone else's research. She had ideas about how to do things differently, but she did not feel that they were valued. Expressing creative ideas was not part of her job. So, she decided to go to graduate school.

She applied for and received a GEM Fellowship,[1] which she had heard about as an undergraduate—and one of her best friends had received one. A working, single parent, she was an unusual selection; most candidates came directly from undergraduate school. Sponsored by DuPont, she did not have to do the required summer internship—and she did not want to move her children to Midland for three months and again to California. As an alternative, she wrote a technical paper based on the work she was doing at Amoco. When she received the fellowship, a lot of information about universities she could attend came to her mother's home. Her mother leaned toward attendance at UC Berkeley, because she wanted to visit California. When Dianne got the application and read about UC Berkeley, she discovered it was the number one public graduate school for environmental engineering at the time. This prompted her to attend UC Berkeley to study environmental engineering. She loved the diversity there. She felt accepted as an intelligent woman. An environmental engineer major with a strong chemistry background, her research project focused on the chemical/biological treatment of municipal waste water to remove phosphorous. At the time, UC Berkeley operated the Richmond Fields station, a remote satellite laboratory. Dianne set up a laboratory-scale waste treatment plant there to study the chemistry of adding aluminum phosphate to wastewater to reduce phosphorous concentration in the effluent. Removing the mineral phosphorus from waste water is important because it can cause eutrophication or algae blooms in the water bodies into which the treated waste water is discharged—this is an issue in any wastewater treatment facility. Her PhD advisors were Professors David Jenkins and Slav Terminotics who managed a research group of about six students doing graduate research.

Dianne lived at Albany Village, a family student housing complex for married or single parent students in the Berkeley suburb of Albany, for the seven years it took to complete her master's degree and PhD. Two years into graduate school, Dianne met her current husband, electrical engineer Tikisa Anderson, at a minority engineering conference in Oakland, CA. They found that they had quite a bit in common. Both were single parents, engineers with graduate degrees, and community volunteers. Much to her relief, Tikisa has always been supportive of her career, even when it involved extensive travel.

After receiving her PhD, Dianne's first job was at Oak Ridge National Laboratory (ORNL) in Oak Ridge, TN, where she worked from 1992 to1995 in the Environmental Science Division, researching the removal of mercury from contaminated waste and in situ treatment of soils contaminated with dense non-aqueous phase liquids (DNAPL) such as trichloroethylene (TCE) and tetrachloroethylene (PCE). She was developing and reviewing technologies to treat waste in place, without having to dig it up and treat it off site.

National laboratory projects are primarily federally funded research managed for the Department of Energy. Different national laboratories might have different focuses or different areas of expertise. The system is flexible, as a PI can look for funding for their own projects or support colleagues on their projects. At ORNL and later at Lawrence Livermore National Laboratory (Livermore), Dianne did both, but her work always involved different types of waste: hazardous waste, mixed waste, or radioactive waste.

After a year at ORNL, she was assigned a mentor, Stan A. R. Baugh, a retired former division lead. New at the lab and unclear about what to expect or how to take the maximum advantage of the opportunity, Dianne thought this was because she might not be working well. As a former division head, Stan had a lot of connections. He helped her to go to conferences, told her what committees to join, and suggested things he thought would be good for her developmentally. She was able to find the funding to follow his suggestions and also found she liked outreach programs to students, which did not count toward her development, according to Stan. She served as a visiting professor for the Urban League HBCU program and was very active in her division's education outreach programs. She served as a one-on-one mentor for a high school student for four years and mentored students in the Expanding Your Horizons program, designed to expose girls from fourth to eighth grades to technical careers and give them an opportunity to interact with women working in technical and other professions. Once a year, on Expand Your Horizons Day, the girls came in to the lab for half a day and rotated through four or five workshops to learn about different aspects of the research and participated in hands-on activities. Dianne preferred eighth-grade students possibly because they were more inquisitive. Being a single parent with job where travel was expected, was a particular challenge. As being a parent was her number one priority, Dianne took creative measures to find good caregivers for her children so she could travel. When she lived in Knoxville, she went to the small HBCU there and asked one of the staff members for a recommendation. She needed a very responsible student who could do live-in caregiving on an irregular basis while she traveled. It worked! She found someone who could stay with the kids. In the summer, she would "rent" one of her bedrooms to a summer student in exchange for childcare when she traveled. She tried to limit her time away to no longer than three or four days on any one trip.

In 1994, Tikisa and Dianne married and decided to move back to California. ORNL did not want her to leave and Dianne found getting another job at a national laboratory more difficult than she had expected. But she had good letters of references from the division director and the lab director, plus a letter of introduction to the chair of the environmental department at Livermore. The perception of prestige and elitism in the lab was a particular barrier when she applied there. Even though she had worked at ORNL on environmental science, at Livermore there was a different focus. She visited, gave a seminar, talked to people, shook the right hands, but still did not get a job offer. She looked for work elsewhere, applying for a job as associate director with the Minority Engineering and Science Academy (MESA), run by UC to increase the number of minority students in engineering and science. After a half-day interview in front of the board, she learned they already knew who they wanted and had only posted the opening to meet legal requirements. She was disappointed; but, one of the people on the review board was Karl Pister, then chancellor of UC Santa Cruz. He had been dean of engineering at UC Berkeley when Dianne attended, and they had met when students wanted to diversify the faculty. He called her and offered to help her because he was on a review board for Livermore. He took her resume to the lab director, or to someone really high in the organization, and Dianne got an interview and was received a job offer. There were not many minority women at Livermore when she began and there are not many now, almost twenty years later.

At Livermore, she worked in the Radioactive and Hazardous Waste Management Division of the Waste Treatment Group, developing processes to treat unique and hard to manage waste. She was the principal investigator on a three-year Lab Directed Research and Development Project (LDRD) and a co-PI or major contributor on a variety of other projects. Later, she worked on "consequence management," helping urban communities respond to terrorist events, like cleaning up after a dirty bomb is exploded.

Dianne has one patent for high-efficiency particulate air (HEPA) filter encapsulation.[2] When working with radioactive waste, HEPA filters are installed in buildings to capture particulate matter; in particular, lab flues are connected to such filters, which may become contaminated (by the waste in the air passing through them) so they cannot be disposed of in the usual way. Waste disposal facilities have very stringent rules on the kind of waste they can accept and in what form. In the United States, low-level radioactive waste goes to the Nevada test site disposal site, which doesn't accept mixed waste (containing both radioactive or hazardous waste). Dianne's project team developed a process that would treat the hazardous filter components so it could be reclassified as only radioactive waste and could then be more easily disposed of as it would be acceptable at the Nevada test site.

Dianne retired from Lawrence Livermore National Laboratory in December 2016 after nineteen years of service.

When asked how to encourage girls and minorities in science, she commented that STEM careers have been emphasized for some time. Currently there seem to be more opportunities and more activities to engage girls in these careers. She thinks we must continue to wholeheartedly support such activities and endeavors, which requires a lot of work and energy. In California, Dianne worked every Saturday for six weeks in a row at the Saturday Science Academy sponsored by NOBCChE.

Would Dianne encourage minorities to go into science? Yes, she would encourage them to pursue it. For anyone with an interest in science and technology and the desire to work in the field, it's a good career. But it's really important to start preparing early for a STEM career. Public schools need to make a better effort to provide a solid academic foundation for all students.

Dianne's family proves she has practiced what she preaches. Her children are now grown. Her daughter is an electrical/computer engineer and computer coder living in Denver, CO; her son is an auto mechanic; and her step-son is an electrical engineer. She also has an adult step-daughter who is developmentally disabled and three grandchildren. Since Dianne has recently retired she has not yet decided what to do in her free time, but she would like to help kids in science. She's looking into where she would be needed most.[3]

5.3 Allison Ann Aldridge

FIGURE 5.3. Allison Ann Aldridge

Chemist Allison Aldridge (Fig. 5.3) was born in Kansas to a military family and worked in numerous chemical labs doing analytical chemistry before working for the FDA, despite being born with a disability. She shows that anyone can be a chemist despite having a disability.

Allison was born on October 11, 1961, in Ottawa, KA. Her father was in the military so they moved around a lot. She had three brothers, one older and two younger. The youngest brother was from a different mother than hers. Since her father was in the Air Force, he was deployed in many areas, but the family's home base was Ottawa, as both her parents were born there. Ottawa was a small town, and her grandparents were neighbors before they married and had children. Her mother's parents came to Ottawa from Cherryvale, KA, and her father's parents from Lawrence, KA. One of her grandmothers lived near Allison's family in Ottawa and would take care of her when her mother was working. Her mother started to go to college, but left to get married and have kids.

Allison lived in Ottawa while she was in the third and until the seventh grade, because her father was deployed to places, including Thailand, where the family could not join him. Since Ottawa was a small town, there was only one high school and one junior high school, but four elementary schools. There were very few African Americans in the town, so, when she got lost, she was easily identified and returned home. And since it was a small town and there was only one high school, the school was integrated. One white girl was afraid that if she touched Allison, she would turn black. Allison did touch her but she was not kicked out of the school even though the teachers and parents may have wanted to.

Allison had no science classes in elementary school, her first was a chemistry class in high school. She decided to be a veterinarian because her grandfather had pigs and took high school chemistry as a precursor to veterinary studies. She used to name the pigs, and when they were butchered she refused to eat the runt because he was her favorite. She did not want to be a doctor because she thought she would become too emotionally attached to her patients. She then thought she might be a pathologist because her patients would all be dead, but she thought she might get tired of that. Still, she stuck with that idea until she went to college. She took business and accounting classes in high school, but she thought there would be no money in that profession so she decided to continue with science. She also did not like the business courses. She was in the National Junior Honor Society and the National Honor Society.

Allison applied to only one college, the University of Illinois Urbana-Champaign, and was accepted. She does not recommend only applying to one college, because what would she have done if she had not been accepted? When she entered college in 1979, she was seventeen. Her plan was to get a degree and then go to the vet school there. But when she took undergraduate research, she wound up working with chickens, because farm students were preferred for working with larger animals. Although majoring in biology, she took several chemistry courses. She thought about switching to an Animal Science major, but she would not have been able to do the research because of her disability.

For financial reasons, Allison went to the University of Kansas for a year to complete her degree. She wanted a bachelor's degree in chemistry but lacked core courses, like English and Western Civilization; she earned a bachelor of science in biology from University of Illinois Urbana-Champaign in 1983.

Allison got a job as a lab tech in a quality assurance lab for Hercules Aerospace, Inc., where both her parents worked. She would have to go out where she was told to and get the samples and bring them to the lab and test them. Since Hercules was an explosives company sometimes did have small explosions. The only requirement for the job was being able to lift fifty pounds. She learned early in life to ask for help if she needed to do something she could not do, though she also learned to do some things herself. She asked the guys to lift heavy objects for her. One guy objected, but she said she would do the things in the lab that the guys did not want to do, so they would be even. She enjoyed working as a lab tech, although when she started she did not think she was doing chemistry. She also kept asking questions in the lab and an African American woman working there became her mentor. She would help Allison and keep tabs on her, and she encouraged her to look for other jobs or get an advanced degree. Almost all the people working in the lab were chemists even though it was not a requirement to do the job. As Allison kept questioning things, she was told to study more so she could go for a PhD. She starting looking for other jobs but was losing them to chemists with advanced degrees. So, she decided to take courses so she could go to grad school, which was possible because she worked on flex time. Her colleagues at the lab helped her while she was studying. For one course she had to get a tutor to help her. She was working and going to school and sometimes she had to miss school because of work. She was told by one professor she should chose school or work, but she could not do both.[4] But, the lab professor was good to her and helped her when he could. But it turns she did not drop analytical chemistry and on her record, they gave her an F. That came back to haunt her when she applied to graduate school at Loyola University in Chicago and was asked why she wanted a PhD in analytical chemistry when she had flunked the course. She told them that she had been working as an analytical chemist for eight years! They let her enroll on probation.

Allison had chosen Loyola for grad school because the faculty was diverse, unlike the Illinois Institute of Technology, which she had considered. But when she visited, she did not feel comfortable there, because she was African American. In choosing between graduate schools she was also trying to decide between two degrees, environmental chemistry and environmental engineering. She applied to Northwestern University and was invited to visit; but after they saw her record they disinvited her because she had no engineering courses. When she looked at the photo of the faculty, she saw that the faculty was not diverse. At Loyola, she was welcomed with open arms. The chairman of the department greeted her and gave her a tour, and she met Dr. Willetta Green-Johnson—the first African American chemistry professor that she had ever met.

She didn't start at Loyola until 1993, although she was admitted before, as she was hoping to be laid off from work; the company was dropping a lot of people and a lay-off would mean more money in her discharge packet. But eventually she decided to resign so she could go to college. Her plan was to get a PhD, but in case she got tired of studying, she would do the course work for a master's degree.

She took the regular classes, had a teaching assistantship, and received an Illinois Minority Graduate Incentive Program Fellowship. She also started her research for her dissertation, which consists of two parts because she worked with two different professors. The first part was an NMR study of hydroxyl piperazine ethane sulfonic acid; the other was quantitation of DNA adducts using MALDI (matrix-assisted laser desorption/ionization). For this second part of her thesis she worked with a professor who was brand new to Loyola and had four graduate students. The project was comparing placentas of women who smoked with those of women who did not smoke. The students extracted DNA and built an instrument to study peptides such as lysine and hemoglobin and other chemicals. This was the fun part of the work. But working with the professor was hard, because she and the rest of the team could not get along with him. Most of the team members left his group. Allison continued because she wanted a publication. (But she also secretly resumed her other project with another professor.) Still her relationship with the new professor got worse; he didn't like her because the other students would ask her questions instead of him, since she was an experienced chemist and easier to talk to. So, eventually she chose to leave his group. In spite of this problem she did get the publication.[5] When she finished her PhD, Allison considered working in academia so she wrote a research proposal and sent it to colleges and universities. She kept getting rejection letters, she thinks maybe because the proposal wasn't good enough. (ACS now teaches courses on how to write a research proposal.)

Allison gave a talk at student night at the Society for Applied Spectroscopy. Some employees from Unilever HPC USA were there and liked her talk and offered her a job. She was also recommended for the job because she was a mass spectroscopist. She decided to take the job, because they had great benefits and they gave her three weeks' vacation.

She worked there from 1997 to 2000. The problem was her boss; many people could not work with him, so they quit. Allison learned a lot from him even though he did not have a good temperament. But she started looking for other jobs in the Chicago area and looked at ads in *C&E News*, where she found an ad for Abbott Labs in Chicago. She applied for a job and was granted an interview. After waiting a while she got a call that she had gotten the job, and she took it, even though she was unable to negotiate the terms that she wanted, to get away from her boss. She was hired as a Senior Research Development Scientist in Analytical Chemistry. After moving around in the company doing many types of jobs she finally was placed in one she liked. Since Abbott was a pharmaceutical company they worked with chemicals that had the potential to become new drugs. At first Allison worked on late-stage drugs—those that will be going onto the market. She had to make sure that all the test results for those drugs were accurate so that they could be passed on to other labs. That was hard because she was not doing anything new. Then she got transferred to early-stage drugs, otherwise known as discovery. Here she was able to develop the tests for the drugs. They started her on a project already in process. A colleague was working on step one of the testing, and Allison was supposed to work on raw materials and then step two. But she looked at what her colleague was doing on step one and decided she could do it better and did, which hurt her coworker's feelings.

Allison joined some affinity groups at Abbott, such as the Black Business Network and Women's Leadership Action. Many senior managers were in the Black Business Network, and a lot of the executives were on the board. Then they asked you to join, if you want to join. Membership gave her the opportunity to meet and learn from many of the successful people that had been with the company for a long time, including other African Americans. They give seminars. They talk to you one-on-one informally and they also had speakers, that sort of thing. The other group was the Women's Leadership Network which was similar. Women executives were on its board, and Allison was able to work with a lot of successful women who had a lot of wisdom to impart. She heard a talk given by Melody Hobson—president of Ariel Investments—on the mistakes she had made early in her career. Allison found that very insightful and inspirational and thought it took a lot of courage to talk publically about career mistakes to help other people avoid them on their way through their corporate career. These groups helped Allison learn how to get along in corporate America, and she would encourage young people who work for corporate America to join an affinity group, if one is available. She was promoted once and then tried for an in-house promotion at Abbott but she was not selected.

She felt that in industry, for the most part, with only a bachelor's degree, she was not allowed to think, they treated her as just a pair of hands. But one thing she did do was to cheer up her co-workers if they felt they had been mistreated. They would come to her and she would cheer them up.

Allison decided after she had been at Abbott for five years that it was time to move on to another company. She said that she comes from being part of a generation that wanted to keep moving. In order to do that, she recommends that people keep their resume fresh. As you learn a new skill, you should add it to your resume. Abbott labs was beginning to downsize and she decided to get out before the real downsizing happened. She started her job search by sending letters to corporations listed in *C&E News* and also searched company websites, Monster.com, and Career Builder. She kept all the letters and got a call from a small contract pharmaceutical company called Mikart in Georgia and went for the interview because she wanted to get out of the cold weather in the north. After she had accepted the position there, she got a call from DuPont, in Philadelphia.

Allison started work at Mikart in 2005, as a Senior Scientist Supervisor and was subsequently promoted to supervisor and then Manager of Analytical Services. They had some new equipment UHPLC (Ultra High Performance Liquid Chromatography), she was excited about working with this new equipment. She had to develop methods and train the analyst to do the drug testing. One of her responsibilities was verifying that a testing procedure would work. Five people read the reports and signed off on them. Sometimes Allison would find an error that other people missed. This was important because this information needed to be sent to the FDA so she would have to tell her boss about the error.

Allison left Mikart to take a position at Revogenex Incorporated, a brand-new company (a start-up), as Manager of Analytical Services in the R&D lab and quality control. Revogenex is a pharmaceutical drug development company focusing on drugs for irritable bowel problems. The company was considering opening a division in China, which might have given Allison a chance to travel. Revogenex does not plan to manufacture commercial drugs, just to develop them and make clinical batches.

While she was there, they were working on drugs that were already on the market but never formally approved by the USA. Many drugs go on the market in other companies before they are approved by the United States. She liked that job a lot, because as it was a small business, she got to learn a lot about the company. On the other hand, in a large company money was no object, but in a small business money counts, so she had to learn to make do with what she had.

In 2009, Allison changed jobs again to become Director of Analytical Services at Speed Laboratory, Inc., in Buford, GA, where she worked until 2011. In January 2011, she joined the FDA as a Chemist in the Office of Compliance at Center for Drug Evaluation and Research (CDER). Since then, she has been promoted to Team Lead, guiding, mentoring, and coaching others and training new hires. In October 2016, she moved to the Office of Pharmaceutical Quality, where her primary responsibility is reviewing the facilities submitted in drug applications. She also joined a newly formed group, The Women of CDER, and participates in the mentoring program of CDER.

FDA is a scientific regulatory agency that has adopted a more risk-based assessment in its regulatory reviews. This has provided both challenges and opportunities for chemists to rethink the traditional roles of review, enforcement, and policy to support a more collaborative inter-departmental function.

Allison says that: "Women chemists and those with advanced degrees in particular have performed a critical role in formulating, testing, reviewing, and appraising product quality. At FDA, women chemists work at all levels, ensuring review of drugs for safety and effectiveness to help protect the public health."

Because of her work commitments, Allison has had no time to work with local kids doing science. But she did have time to work with the national ACS, which she joined 1996. She is a member of the Committee on Chemists with Disabilities and a career consultant; and was secretary of the Professional Relations Division, 2004–2009; member of the Committee on Minority Affairs (CMA), 2008–2012, its chair from 2009 to 2011. While she was chair, the committee was working on programs to engage more underrepresented chemists and students and bring them into the ACS. They discussed having joint meetings with NOBCChE, Society for the Advancement of Chicanos/Latinos and Native Americans in Science (SACNAS), and American Indian Science and Engineering Society

(AISES). Allison's goal was to make those joint meetings happen. She also wanted more minority scientists involved in the work of the ACS. While in the Chicago local section she was Alternate Councilor for the ACS from 2001 to 2003, Councilor from 2004 to 2005, and Chair-Elect in 2005.

Allison is now a member of the Chemical Society of Washington (local section for the metropolitan area). She is currently a councilor and president-elect and serves as the local Project SEED coordinator.

Her advice for minority chemists is to establish or join a network. Find a mentor to help. Allison had a mentor but she did not know that the woman that encouraged her was called a mentor. She said there is a lot of things you can learn from other people and they can learn from you.

More advice: If choosing to go into corporate America, try small companies, because it's a whole different thing, requiring different skills than in a large corporation. So, she thinks both have advantages to someone's career. The more experience a person has the easier it would be to choose a company. Allison says that her second career may be teaching. She likes to work at science fairs with kids, to answer their questions and do hands-on science with them. She says you never know where the next scientist will be.

Here is what Allison would say to a young person who wants to be a scientist: "you have to have a passion for it. If you don't have a passion for it, it's not really worth it because it's hard enough just working every day, but if you hate it or don't have a passion for it, you'll be ready to quit after a couple weeks." If something really excites a person, they should figure out a way to make a living at it.

Allison likes talking to kids because she wants to encourage them to become scientists and ask questions. She said she might not have become a scientist had she not been encouraged to do so.

Here is what she would say to a college student: "you don't have to be brilliant to get a Ph.D. You'd have to work . . . The truth about PhDs, we know a whole lot about one little thing. But, we know where to get information from. . . . the greatest thing about getting a PhD is learning how to think, learning how to ask questions. You answer your question; it should lead you to ten more."

She once spoke to a nineteen-year-old student who was surprised that Allison had a PhD because she looked so young. She said you don't have to be old or white, to get a PhD.[6]

5.4 LaTonya Mitchell-Holmes

FIGURE 5.4. LaTonya Mitchell-Holmes

LaTonya Mitchell-Holmes (Fig. 5.4) is the first female African American District Director of the Food and Drug Administration (FDA) Denver District Office.

LaTonya's maternal grandmother Bessie White played a significant role in her childhood. LaTonya spent many summers with Bessie who told her about the importance of obtaining good grades in school and encouraged her to read and use proper grammar when she spoke. Her grandmother was a caregiver, who had worked in California for an actress, taking care of her children, and then moved back to Arkansas to take care of her mother.

LaTonya was born in Kansas City, KA, at the University of Kansas Medical Center on April 28, 1969. Her mother Mary Ann Wilson works in the administrative field as a secretary and grew up in Pine Bluff, AK. Mary Ann graduated from Merrill High School in 1965 and went to Arkansas Agricultural, Mechanical & Normal College, which is now the University of Arkansas, Pine Bluff, but did not get a college degree. LaTonya's father Jesse Mitchell is a general contractor, the owner of Mitchell's Contracting, and works at Fort Leavenworth. He was born in Tupelo, MI, and moved to Kansas City in 1953. He graduated from Sumner high school in 1965. Jesse is an army veteran and served in the Vietnam War. LaTonya's parents, both still working, met in Kansas City. She has one brother, Eric, who is three years older, and he and LaTonya's mother currently live in Pine Bluff.

LaTonya went to school in Kansas City, in Wyandotte County, District 500 School District, where she received a very good public education. She attended Parker Elementary school for kindergarten but then her family moved to the suburbs, to a predominantly white neighborhood—even though Kansas City was integrated, this was a white neighborhood. From the first to the sixth grade, LaTonya attended White Church Elementary School and was the only African American girl in her class—although there were two African American girls in the grade, the other African American girl, Sherry, was in a different class. LaTonya enjoyed going to school, and in spite of being in an all-white school, she made a lot of friends. In seventh grade, she went for one year to Eisenhower Junior High School, now Eisenhower Middle School. When LaTonya was there, it was more integrated and this was her first time being around a lot of black students. LaTonya took science in junior high school and she loved it. She was always drawn to math and science and really did not like English.

From eighth through twelfth grade, LaTonya went to Summer Academy of Arts and Science, a magnet school to which students must be accepted. She really did not want to go there, but she was encouraged by her parents. In the end, she loved the school. She took chemistry in the eleventh grade in high school and chemistry 2 in the twelfth grade. She had gotten the bug for chemistry early in high school. When she went to college she wanted to major in chemistry. She had a good high school chemistry teacher, Mr. Ell. The way he taught her made her appreciate chemistry. She took other STEM courses at Sumner. She wanted to take a home economics course, like cooking or sewing, but Sumner did not have a home economics curriculum. Her counselor told her that that "home economics is for the home and not for the school." He said parents need to teach you that. High school was a little more integrated but still pre-dominately white. In her graduating class of 1987 there were about one hundred and fifty students with just twenty-three blacks. She graduated from high school in May 1987.

In August 1987, LaTonya went to the University of Kansas but left after one semester because she was expecting a child. Wanting to stay in school, she transferred to Kansas City Community College for the second semester. Her son (her only child) was born in 1988 and is now a young man. She was able to take care of him while still going to school. A single mother, she took all the courses that she needed, science courses like chem 1, and chem 2 and math courses, and she received an associate's degree in science in 1989. But, when she transferred to Park College, now

Park University in Parkville, MI, there were still gaps in her education she was required to fill. At Park, she did work in laboratories as an undergrad. She worked part time and while in school. In the summer, she worked full time at Citibank National Associate as a credit card collector. She mainly worked in telemarketing.

She graduated in 1993, with a bachelor of arts degree in chemistry, with a minor in biology.

LaTonya had planned to go to medical school but then decided on pharmacy school and was accepted into the University of Kansas pharmacy program. She really wanted to work at the FDA and start making some money. But, after she graduated from college it took her a year before she found a job in the science field. Meanwhile, she was working at Citibank, where she had worked while in college. She needed to take two classes to get into the pharmacy program, so in January 1994 she went back to the community college and took those classes before she applied to the pharmacy program. At the same time, she applied to work at the FDA. She applied in November of 1993, was interviewed in March, and started her job May 1, 1994, as a GS 7 (General Schedule 7) because of her academic scholastic achievements as an undergrad and because her GPA (grade point average) was high. So, even though she had been accepted into the University of Kansas pharmacy program, she decided to decline and work at the FDA because she really wanted to work as a chemist.

The FDA regulates foods, drugs, medical devices, and cosmetic products. Working there as a pesticide and elemental analysis chemist, LaTonya was in the food chemistry program, analyzing food and woodware samples to determine if there were any pesticide residues in the product and that these complied with the tolerances set by the Environmental Protection Agency (EPA). She did this job for seven years, and in her eighth year set up the quality management system. The FDA was trying to get their laboratories accredited by the American Association for Laboratory Accreditation (A2LA), for which they needed to put in place a quality management system with policies and procedures, they succeeded in getting the labs accredited.

LaTonya was able to get a master's degree in health services administration from Central Michigan University in 2002 by taking classes in a satellite of the university in Kansas City. Her classmates were a diverse group, with from ten to fifteen in the class, men and women, blacks and whites, including a lot of military students. She did not take any labs for her master's degree; her thesis was based on the development of a quality management system for the FDA. She was happy because the FDA paid for half of her classes—she was working full time and going to school part time in the evenings. Her director told her if she took the class and received a B or better, he would pay for the next class.

LaTonya also conducted inspections of industry and drug companies regulated by the FDA. A Good Manufacturing Practice Inspection (GMP) was conducted regularly during which methodologies were reviewed and tested for a pre-approval assignment to manufacture the drug. She accompanied investigators on domestic inspection of the labs and manufacturing facilities of the corporations, reviewing the laboratory operation while the investigator reviewed the manufacturing operations. Her inspections included all the methodology, tests, assays, training records, and so. She looked at their instrumentation and did an audit of the laboratory to make sure that they were complying with FDA regulations. She was also accepted to be on the foreign inspection cadre, to accompany a Consumer Safety Officer (CSO) going overseas. But, she became a supervisor and so never did.

In 2002 LaTonya was accepted into FDA's leadership development program and became a supervisor in September of that year. She supervised from ten to twelve chemists, biologists, and technicians, reviewing their analytical packages. She had to deal with the personnel and technical side by reviewing their work. She did that from 2002 until 2006. She wanted to get off the bench but stay in the lab, so she moved to management. Her team was primarily white—when she came to the FDA she was possibly the only African American female chemist. There was a female African American biologist and other female African Americans who were technicians. Technicians don't have science degrees; they may have had some college classes and must have taken at least fifteen hours of science courses. To be classified as a chemist requires a degree or thirty hours of chemistry. This is not a career for young people just starting because the FDA does not hire too many technicians, they want to hire chemists and individuals with a science degree. LaTonya was the only female African American chemists for several years but there are more there now. There was also an African American male chemist who was also first, but there are more now.

LaTonya is now pursuing a doctoral degree in health services administration with a concentration in leadership at Walden University. She has finished taking her core courses and has taken her first research class, introduction to research, which covered theory and qualitative analysis verses quantitative research. This doctoral program is online, which is good for her. She travels with her laptop and books and at night, wherever she is, she reads online or studies. In order to do this, she must manage her time and have good managements skills. Assignments are due every week and there is so much to read and post. She has to have good written communication skills, because there is a lot of writing. She is now thinking about the topic for her dissertation, perhaps recreational marijuana and its impacts as she is now working in Colorado.

LaTonya married Emmanuel Holmes Sr. the father of her son. Her son graduated from the University of Kansas, and her husband is home in Kansas. In 2006, LaTonya moved to Atlanta, GA, as the chemistry branch director at the Southeast Regional Laboratory. Since her husband has his own lawn care business, Person Touch Lawn Service, he stayed home and she commuted from Georgia. She is the only one in her immediate family to have a master's degree; her son has a bachelor's degree in sports management with a minor in business.

When she was chemistry branch director, three supervisors reported directly to her and there were about thirty to thirty-five people in her group. She oversaw the pesticides, food and color additives, and mycotoxins programs. Her team analyzed food, feed products, and grain products for toxins, like aflatoxin, fumonasin, and so on, ensuring they did not exceed allowable levels.

LaTonya thinks she was promoted because of the leadership development program that she took at the FDA. It was a very competitive program, for which they selected from twelve to fifteen employees across the FDA (at GS-12 to GS-14) and prepared them for future leadership opportunities. They did not promise or guarantee them anything, but all the training and all the resources expended pretty much helped her get a foot in the door of a management position. LaTonya was a GS-12 and the only GS-12 they accepted in the program that year. Chemists can start at GS-5 and end at GS-12. A supervisor starts as a GS-13. Each rating has a different pay scale. As branch director, she was a GS-14, a mid-level manager. Currently she is a district director, GS-15 and the highest on the GS scale. The next level is the senior executive service.

LaTonya moved to Denver, CO, to become District Director in 2011 and considers this the most rewarding position that she has had in the FDA. She oversees investigation and compliance work for the human and animal food programs in the five states of Colorado, New Mexico, Utah, Wyoming, and Arizona. Her staff regulate the human and animal food firms in that five-state area; there is an investigation branch and a compliance branch and an administrative unit. When she first got there, they had a lab so the staff was about one hundred and twenty or one hundred and thirty. But in 2014 the labs broke off from the districts. Now each lab is separate and do not fall under the district. So, she just manages and oversees the investigation and compliance work. Her investigators have degrees in chemistry, biology, and pharmacy—they must have a science degree. They all have GS levels starting at 5 and going up to 14. The inspections can be unannounced or announced so that the company can prepare, depending on the type of inspection. Her staff, when at total capacity, would be close to ninety. Her work consists of overseeing the district operations; she holds regular meetings with her staff; oversees the administrative aspect of resource management and personnel management; meets with key leaders for other FDAs organizations and some of her state counterparts; and she is the voice of the FDA in her area, for example in meetings with other government agencies, she met with the HHS (Health and Human Services) and represented the FDA at that meeting. These groups have compliance officers they have the Office of Chief Counsel which is at headquarters and each center has the office of compliance. For this reason, there are lot of attorneys work for the FDA as well; some attorneys have a science degree—LaTonya met with a woman who is a JD/PharmD which means she has both a pharmacy degree and a juris degree.

Although LaTonya's husband remains in Kansas City, he is very supportive of her trying to improve her career. If she had stayed in Kansas City she might have remained a chemist but could have only moved up to the GS-14 position. By moving around, she has been promoted from GS-13 to GS-15 in five years. She does not think her husband resents the fact that she has a higher degree than he has and maybe makes more money than he does. When she is home she knows her role as wife and mother. He is even supportive of the fact that she is working on her PhD. Her husband has an associate's degree. He makes good money with his business and supports the home. She says she does not need her husband to be at the same level as she is, she married for love.

LaTonya thinks that she was promoted because she could do the job. She also says she was very motivated, ambitious, and sought opportunities that would propel her to the next level. She undertook various assignments before she became a supervisor so she could become a supervisor. Before she became the branch director in Atlanta, she would act for her supervisor and then she participated in the leadership program, so, she became qualified for leadership positions. She says that since she is an African American woman she thinks that she has to do twice as much to get half as far. Therefore, she wants to make sure that she has the credentials and qualifications for the positions that she applies for. She did not become a supervisor the first time she applied, but a man with thirty years' experience got the job—she had only six or seven years' experience. She applied to be either the chemistry director in Denver or lab director in Atlanta and received the Atlanta position. When she applied to become a District Director she got the job. She is able to apply internally when the jobs are posted.

LaTonya says she is working on her PhD because she needs to have those credentials to become a senior executive, because the FDA is science based. They want your education to fit the job.

Her mentors were very supportive of her. When LaTonya started at the FDA, her son was in the first grade, and Michael Rogers, the district director, said that when her son graduated, if she wanted to move up she needed to be mobile. He gave her the life lesson that allowed her to maneuver as an African American woman in a predominately white organization. In Atlanta, she had a lot of black male mentors, one was Malcolm Frazier, and she also had two ladies who would give her some direction. They were both official mentors and unofficial mentors. She even has mentors now who are in an equivalent grade, but she can seek advice from them.

LaTonya took a lot of courses within the FDA, which were relevant for them. The FDA does invest in their employees. She also gets a lot of assignments. If there is a need for someone to work in another area for a month or two she would do that. For example, she knew she wanted to become a district director, so when she was in Atlanta there was an opportunity to serve as the acting district director in San Francisco. So, she went there and worked in that position for two months; she also served as the acting district director in Atlanta. As the acting district director, the staff knew that she was there temporarily and they still reported to her. She said she was just trying to keep the status quo and help the staff in the transitional period. But she was building the leadership skillsets to prepare her for the job. This helped her when she interviewed for the job, because she could talk about the things she had already done. The people who reported to her would give her feedback about how she did her job. She even worked as acting laboratory director in Detroit. There they were moving from one building to another, so she had to work with venders to help move the instruments from one location to the next and make sure it was stable. She learned about facility management in that position.

She even followed one of her mentors, Mildred Barber, to San Juan in Puerto Rico to work as a compliance officer for her. She thought she needed to enhance her knowledge in investigation and compliance. Her mentor was district director, but she had also started out as a chemist and had been a supervisor and lab director. She was able to help LaTonya to get where she is today.

LaTonya has numerous awards from the FDA—they do a great job of honoring their employees, with honor awards that are not monetary and incentive award that are. But now she sees that her employees get money to increase their morale.

LaTonya is a member of the Association of Food and Drug Officials (AFDO), which runs educational conferences drawing attendees from state and federal agencies and industry to meet and discuss what's going on at the FDA and United Stated Department of Agriculture (USDA); there are separate tracks for food and for drugs and devices. There are also a lot of educational speeches and training courses, and it is a good opportunity to interact with state and industry counterparts in a none adversarial, collegial setting. She is also on the International and Government Relations Committee (IGRC), which has representatives from Mexico, Canada and the United States, and on the Drug and Device Committee.

LaTonya has been a member of AKA 1997 and was treasurer, financial secretary, a hostess, and a custodian. AKA has collaborated with the National Association for Mental Illness (NAMI) to increase awareness of mental illness and also collaborated with the American Heart Association— these are diseases that have impacted the African American community. Education is one of the AKA's main platforms, so many of the local chapters work with high school and middle school students to get them prepared for college, taking the students on college tours, running college prep programs, and providing scholarships. The organization usually has five programs that they focus on. They have gathered backpacks and put food in them to prevent childhood hunger. They also go to the playgrounds in the minority communities to clean them up so that the children have a safe place to play.

In Denver, LaTonya's life consists mainly of work and study for her PhD, but since 2004, she has been a member of the Oak Ridge Missionary Baptist Church and before that she was a member of Pleasant Green Missionary Baptist Church. She travels back and forth to Kansas City and goes to church in both cities.

Her son is now store manager at Sherwin-Williams (Paint and Coating Manufacturing Company). LaTonya would like him to get his master's degree while he is still young. She was able to raise him because she has a strong supportive system and his father is with him. She was able to keep him busy and active doing different thing so he did not have idle time. This is why she did not leave Kansas City until he went to college. Because of her parents, her husband, and faith, and church he was exposed to good opportunities so that he would not be easily influenced by peer pressure. He did stay focused. His father would coach him in football, basketball, and baseball so he was there throughout his childhood.

LaTonya's contributions to the field include supporting her researchers in their scientific endeavors and research. She reviewed their manuscripts and critiqued their work, asking questions. Her researchers would present their research at a pesticide conference or to the American Association of Analytical Chemists (AOAC), and to her staff. Before the manuscript would go out they would have to have the AOAC evaluate the paper in order to see if it were in good order. LaTonya is a former member of AOAC and of the ACS.

She likes working for the FDA because of its mission, which is to protect and promote public health. She likes knowing that her work is important, keeping consumers aware of the safety of the food and medical products they use. She likes having a piece in the mission to promote and protect public health.

She would tell a young scientist who wants to come to the FDA to learn your job and do the best you can. Don't try to move up too quickly. First make a positive name for yourself. FDA is a great place to work and there are many opportunities, so by biding their time and doing a good job, young people can have a rich and rewarding career. Since LaTonya is a pioneer African American manager, she would say to young women who would like to follow her, to do well in college, make good grades and stay focused. Since it is difficult to get into the FDA, prospective employees must do things that make them stand out, like having the highest scholastic achievement. Don't wait to get a master's degree, or a PhD—these will give you an advantage. When she joined the FDA with just a bachelor's degree no one wanted to work for the government, but now everyone wants to work for the government because of its stability. Young students also need to be tech savvy because of all the computer applications.

LaTonya's short-term and long-term goals are to do a good job as director in Denver and maybe move to Kansas City to become a director there, so she could go home. Long term, she wants to finish her classes for the PhD and write and defend the dissertation.

She has regular speaking engagement each year. In the past, she was asked to speak at an ACS meeting. She provides an FDA update every year to the Rocky Mountain Regulatory Affairs Society. She gave opening remarks at the REdI (FDA regulatory education for Industry) conference. At the Rocky Mountain Food Conference, she will have several people from her staff go and talk about the Food Safety Modernization Act that was signed into law in 2011 and will require industry to change some of their practices to ensure compliance. Therefore, they have to go out and educate the stakeholders.

LaTonya is the first African American female district director in the Denver District Office. She has a wonderful staff, including an African American female deputy director in the investigations branch, though her staff is predominately white. The African American staff consists of about six females, two in the administrative section, two in management, and four who work as investigators. The FDA has special emphasis programs to recruit a diverse workforce. LaTonya is constantly recruiting and would like a diverse staff—when different minds come together to solve a problem, you get different ways of solving the problem. Working with people of diverse backgrounds also helps you to appreciate the differences in the lives of others.[7]

5.5 Novella Bridges

FIGURE 5.5. Novella Bridges

Novella Bridges (Fig. 5.5) is a Senior Research Specialist at the Pacific Northwest Laboratory in Richland, WA. She has been on loan to the Department of Homeland Security in Washington, DC, as a Project Manager. She currently serves as a Project Manager with the Department of Energy, National Nuclear Security Administration.

Novella was born on August 9, 1972, and is the youngest child in her family. Her father is Willie Bridges Jr. and her mother is Carrie H. Bridges. Her mother had four other children from a previous marriage prior to Novella's birth. They are her oldest brother, Lawrence Walker, deceased, Katrina Lewis, Marilyn Braziel, and William King. They were all born and raised in Detroit, MI. Her father was born in Birmingham, AL, and her mother was born in Ocala, FL. Both are now deceased. Her father was the oldest of nine children; her mother was an only child.

When her father was seventeen years old, his family moved to Detroit. As her grandfather was a carpenter and her father had learned the trade, they hoped to take advantage of all the new building of plants and housing facilities going on there at the time. Novella's maternal grandmother, Plessy, then widowed, moved her family from Ocala to Saginaw, MI, in about 1947 to be near her eldest brother and later married Louis "Buster" Heath, who lived in Detroit. So, they all moved again to Detroit, where Novella's mother went to high school, then married, had four children, and became a widow. Novella's father fought in Korea and when he returned, began working as a carpenter with his father. Willie and Carrie met in 1962, through a mutual friend. They fell in love and married in 1965, and Novella was born several years later. Because her older half-siblings were married and had children by then, Novella is younger than her nieces and nephews.

Novella went to a private school, Bethany Lutheran School, an elementary-middle school, on the east side of Detroit and to Lutheran High School East. At the time schools in Detroit were in flux, experimenting with Distributive Education Clubs of America (DECA) and magnet schools. Few had college prep programs. Her parents wanted her to go to college, so they decided not to chance the nontraditional experiments and sent her to a private school with college preparation. They had a choice between the Lutheran school system,[8] the Catholic school system, and the University school system. Both the University of Detroit as well as the University of Mercy had prep schools and have now merged. Since two other private schools were more expensive than the parochial school, they chose the Lutheran schools.

Because the private schools were big on math and science, Novella's science study began in elementary school, as early as the third grade. Detroit had local science fairs in the schools. There were eight or ten Lutheran elementary-middle schools on the east side plus eight or ten on the west side. Each school would hold their own science fair; then there would be an all-schools science fair. Novella always liked science, it was one of her best classes. In elementary school, earth science was her favorite. She had a little pre-chemistry in eighth grade, and, although she liked it, she feels the teacher had poor presentation. Her true love for chemistry began as a sophomore in high school when she had an extremely innovative teacher, Keith Sprow. He brought real life problems into the classroom. Most high schools did not have labs, but her teacher made up hands-on labs for them. Mr. Sprow is still teaching chemistry, but at a different high school, and Novella still contacts him by e-mail to share successes.

Novella graduated from high school in 1990 and applied to go to college. She leaned toward attending Temple University in Philadelphia, but her mother did not want her to go from one inner-city environment to another. There were also funding problems. Novella applied to the HBCUs, which did offer funding, and settled on Jackson State University in Mississippi. She started in August 1990 and found Jackson had an excellent chemistry department and great professors. The chair of the department, Richard Sullivan, PhD, was interested in getting new instrumentation and secured a lot of funding from the NSF. He also worked a lot with the former Upjohn Company, now Pfizer, using his industrial and government connections, he brought in a lot of federal funding. The department had both an NMR and an infrared spectroscopy (IR) machine, which enabled students to use live instrumentation as undergraduates. The courses were ACS certified, because they had a higher level of rigor than some other HBCU programs. Novella was happy that she went from a great high school teacher to really smart and innovative college professors.

Novella did undergraduate research from her sophomore to her senior year with her organic chemistry professor, Dr. Ken Lee. She performed multi-step bioinorganic synthesis and boned up on all the organic reactions, learning all the techniques in the lab from graduate students. She became one of the students who caught on the fastest and credits this to a really good teacher and good graduate students willing to teach her.

Jackson encouraged undergraduate students to get research experience in the field, with internships in corporations. After Novella decided she wanted to do internships in private industry, she found positions with Drackett Company and Kraft Foods. Drackett is a small company in Cincinnati, OH, now owned by SE Johnson Wax. At Drackett she met a man whose wife had a PhD in chemistry and worked for Avon. Since his wife had gone directly to a PhD, he encouraged her to go directly for a PhD in chemistry, telling her she would have better career opportunities with a PhD, instead of studying for a double master's in chemistry and chemical engineering. When Novella spoke about it to her advisor at Jackson State, Dr. Lee, he told her he had always wanted her to get a PhD and was glad others agreed with him. Unlike many of her classmates who were going on to med school, dental school, or optometry school, Novella wanted to do research. She made up her mind about the PhD at the end of her sophomore year and started applying to graduate schools nationwide; she was accepted by all of the schools to which she had applied.

Novella was also good in athletics. She was an academic All-American tennis player. During her senior year, she was named top female athlete at Jackson. For this honor, she attended the National Collegiate Athletic Association (NCAA) for All-American ceremony in New Orleans. The NCAA offers scholarships to encourage college athletes not to go pro but to attend graduate school, and Novella had a chance for a scholarship. In her hunt for funding on a different front, Novella applied for and won a graduate fellowship award sponsored by Entergy Corporation—the power and light company in the southern region—which required attendance at a school in the southern region. When she researched schools in the region, the only school that had the chemistry program she wanted was LSU. Because she was still thinking about also getting a master's in chemical engineering, she needed to find a school that would offer both chemistry and chemical engineering. She had applied for both the GEM and NSF fellowships. When she received the GEM Fellowship, her search was further complicated by the need for a school associated with GEM. LSU was the only candidate that met all her needs at the time. In April, she attended the NCAA awards ceremony in New Orleans and learned she was only first runner up for their scholarship. At that meeting she was able to get a view of life in Baton Rouge. She had already told LSU she would come, and as she did not want to go back to Michigan, this visit helped her to make the final decision to attend LSU.

When she first started at LSU, Novella's thesis advisor was doing extended x-ray adsorption fine structure (EXAFS) and fluorescence spectroscopy in physical chemistry. But they didn't mesh and she thought the relationship was not going to work. So, she switched her major to inorganic chemistry and began searching for another advisor. She interviewed all the members of her committee, found someone who was doing something she was interested in, and began working with Dr. George Stanley. They discovered they were both really good at looking beyond the problem, beyond what the project was, when seeking solutions. He was doing multi-step organometallic synthesis which, because of her undergraduate organic chemistry research with Dr. Lee, intrigued her. She says it was like picking up where she had left off as an undergrad. She was good in the lab and when her advisor would "go off on tangents," she would go into the lab and try to make it work. So, they got along well and, in spite of switching advisors, she was still able to finish in five years.

When asked about other minority chemistry majors in LSU, she said that she came into the department in 1995, after the initial group or fifteen or twenty African-American students recruited by Dr. Isiah Warner.[9] When she arrived, there were about seven or eight minority students in various programs. She and another young woman made the number up to ten, although their group eventually diminished to about four or five. During her time in the program, Novella saw a steady influx of black graduate students totaling thirty or forty graduate students of all races and from many different countries.

However, these students became a commodity. The professors wanted minority students in their group because it would help their programs get grant proposal money. Students needed to know if they were being recruited into a group because they were really wanted for themselves and professors were willing to help them be successful. Novella thinks her problem with her first thesis advisor was because he needed more females in his group. He tried to tell her what kind of person she should be rather than to simply guide and direct her scientific research. He once told her committee: "Well, I'm not really worried that Novella will be a professional woman. She will be, I'm positive, she is well-spoken and she'll be a professional woman. But I just don't think she'll be a scientist."

The committee members reacted by dismissing him from her committee and his assignment as a major advisor. One of her committee members apologized to her. "Miss Bridges," he said, "I don't want his comments to reflect negatively on the rest of us." Later Novella learned that her new advisor was boasting to the former advisor about her lab work, how many interviews she had, and how many jobs offers she received. Her former advisor came to her and said," Wow! You really surpassed what I thought." Her reply was, "Thought is in the mind, but doing is what you really are. The action is behind who you really are."

PhD students are qualified at LSU in the following manner: Students enter the school and declare a PhD or a master's degree track, and all students take qualifying exams by the end of their second year. The exams cover several areas of chemistry—inorganic, organic, physical, and macromolecular or polymers. Depending on the school, the student must pass all five or four out of the five; those who fail the exams can still do research but will only get a master's degree. After qualifying on the exams, students can take further courses and begin original research. They have to write their thoughts for their research project in both an independent proposal and a research update, present them, and, if projects are accepted, become a PhD candidate. The presentations are given both in writing and orally. The first step is a research paper, the proposal must explain the student's proposed method for the hands on, practical research—the how. It must include work in the lab, with an update of what the lab work reveals. Next, the candidate decides what area of chemistry they want to work in and picks their committee members. This involves interviewing professors they might want to work with and selecting three to serve on the committee. The university then selects an independent committee member who is not in chemistry to represent an objective point of view.

Novella's GEM fellowship initially paid for everything at school. But when costs went up, it paid only a portion of her tuition. The Southern Educational Board had a program with LSU to increase the number of African-Americans going to graduate school. The graduate school awarded Novella a fellowship which covered her tuition and GEM paid for her stipend, so she didn't have to become a TA or a research assistant (RA) until her fifth year at school, when the terms of the GEM fellowship ran out.

Novella's PhD thesis was the mechanistic study of a hydroformylation catalyst. Hydroformylation is homogeneous catalysis, a catalytic reactions in which both catalyst and reactants are in the same phase, say liquid or solid. She studied the mechanism pathway of her catalyst. She did the reaction of hydroformylation, which is the process of making a carbonyl (CO) ligand and the ligand detaches or breaks off and makes an alkene (olefin). Her systems were bimetallic so they used a double metal and there were carbonyls around it. When the hydroformylation reaction occurs, you would add a hydrogen and a carbon and the reaction would happen. It was basically an oxidation reaction. When the ligand broke off you would make the alkene that may be six chains or longer. It was a multi-step synthesis. They started with rhodium metals, they bonded the metals together and then the built the catalyst. Then they did the hydroformylation by using an autoclave system where they added hydrogen in a closed system and did the reaction under a pressured autoclave system. The final product would be the alkene. They were trying to produce products that could be used in the oil and surfactant industry. DuPont and also Dow used them to make things like Teflon, surfactants and lubricants.

Novella also became a volunteer teacher. One of her friends from church, a director of the Baton Rouge YWCA branch office, had funding from the Evenflo Company to help parents under the age of twenty-five who had either dropped out or never attended high school and had children while they were teenagers to get their Graduate Equivalent Diploma (GED). Novella was asked to teach mathematics because they could not find anyone who was qualified. Her friend offered to help her become certified as a GED teacher through the East Baton Rouge School District. Novella ended up primarily teaching math and chemistry. The experience helped her see how important math and science are to so many people, and four of her students received their GEDs. Novella speaks proudly of their hard work despite significant life hurdles. The East Baton Rouge Parish school district asked her to come back to teach after graduation from LSU.

After about a year and a half of work with her new research advisor, Novella felt the need to graduate and begin her life. It was time to defend her thesis work. In December 2000, she gave the committee a research update; defended it before the five men, her research advisor, three other committee members and the outside member; answered their questions about everything and left the room while they voted. Each could give her thumbs up or thumbs down and the more thumbs up, the better. Then she had to take the thesis to the graduate school for editing according to the school's guidelines and rules. When the process was completed, she would graduate.

Novella had begun looking for jobs in 1999, a year before she was to defend her thesis, because she knew it would take some time find the job she wanted. She applied for both industry and government jobs, knowing that the government has a slower hiring process. She got interviews when she went to meetings of NOBCChE, ACS, and the National Society of Black Engineers (NSBE). Some would lead to second interviews. She told the interviewers that she had not yet defended her thesis, but the defense date was scheduled. Many companies made her job offers knowing she would be finishing her degree in a matter of months. Since the recession was starting, she applied to several federal government agencies: USDA, FDA, Federal Bureau of Investigation (FBI), Central Intelligence Agency (CIA), and some of the national laboratories. She chose corporations to approach by going through lists of participating organizations maintained by GEM and also some corporations in Michigan where she intended to live. By the time she defended her thesis she had offers from DuPont, Dow, Union Carbide, the USDA, and Pacific Northwest National Laboratory, to name just a few. Because of the recession several companies called to tell her they had a hiring freeze, but Pacific Northwest National Laboratory was still interested. They said they could hire her despite the freeze. Novella is very headstrong, so instead of letting the interviewer tell her what they were going to do, she was proactive. The interviewer told her that they normally hired PhD students as postdocs and then promoted them as scientist. Novella said she would not take the job unless they brought her in as a scientist. The interviewer said she would have to prove herself. She said that was fine but she wanted to brought in as a scientist. They did one better and brought her in as a research scientist (II) because she her PhD.

When she went to the lab in January 2001, she worked very hard, choosing to take the lead on many programs and projects, purposely working on different things. She moved outside her group and its work on catalysis, getting involved in energy systems and even radiochemistry, which was new to her. She learned about (heavy) transition metals and techniques for working with them. Because Pacific Northwest National Laboratory was one of the Manhattan Project locations, there is a legacy of clean-up there, including quite a bit of radiochemistry done at the Department of Energy (DOE) Hanford site. She learned all that and picked up the PUREX process for producing plutonium. She worked on how to remove heavy water from the reactions she was conducting. Because she was good at multi-step synthesis, the lab had her take waste stream products and separate them. She explains that this is possible. If the products are really heavy things, they can be separated using nasty products like cyanide. It becomes a reaction like that of oil mixed with water. Then it's a burping using a separation tube. The heavier products go into a hexane layer and the others remain in the oil products. This is just one method. Her mentor in this area was a nuclear engineer who taught her the nuclear engineering side of the process while she learned the chemistry side. Their expertise complemented each other.

While working in the radiochemistry lab, Novella had to be protected from radiation, using techniques to meet the As Low as Reasonably Achievable (ALARA) standard. To do this, she worked with a shield in front of her and wore double gloves and a special lab coat. She had to make sure that she didn't drop anything because it was radioactive. If she dropped it she would not be protected. There was a Geiger meter near her to check her hands, and if her gloves got hot, she would remove them and put on a fresh set of gloves.

So, during Novella's first couple of years at the lab, she really stretched herself. She was promoted, took on a lot of responsibility, led a lot of projects, and became a PI on different projects. She felt it was important to learn radiochemistry, because scientists at the labs were beginning to get into nuclear medicine development. Because she had the training to handle radio isotopes she was able to partner with a lot of people both in and out of the lab. She had the opportunity to work on some big NIH and NSF grants as PI and co-PI while still very young and early in her career. When most young scientists are being given assignments, she was leading work. Never afraid to be out front or give presentations, Novella knew that she was not going to spend the rest of her life in the lab. She began to venture out. She wanted to be the lead scientist, doing the legwork, and managing other people working on her ideas.

Since Novella first started, there have been changes to the way people work in the labs. Formerly people worked in teams with others in the same discipline. Now interdisciplinary teams are routine. As she looks back on her career, Novella recognizes she was doing that before it became the norm, because she did not want to be one sided. At that time, a PI administered the budget for the team and would give the research scientists a percentage of the budget to complete the project. Novella would normally be given more responsibility and more budget. When the PI was rewriting the proposal, she would ask to become a co-PI. This meant she would no longer be given money as a researcher but would control it as part of the grant. Then she could hire summer interns to work on her project, which brought highly motivated fresh talent to her team and helped her give back by developing other scientists. Because she was working so fast, she was at a higher level than most of the other people who came in as postdocs. Despite her pay grade, she had more financial responsibility than many of the postdocs at the lab. As a result, the lab made her a project manager (PM), responsible for writing proposals to bring in funding.

The national labs are wedged between academia and industry. Sometimes they get their money like industry and sometimes like academia. She was successful at getting funding from the NIH, the NSF and also some private organizations for her radiochemistry, radioisotope medicine project, and a search for a therapeutic cancer agent. At the Hanford lab where she was doing radiochemistry, she was instrumental in getting funding from the DOE, environmental management (EM) for her work. For radiological nuclear projects, the money came from the Department of Homeland Security Customs and Border Protection division. For that job, she was loaned to those agencies to train US Customs and Border officers all over the country.

Job titles at the national labs change according to the job. Her PM title was created when Novella was no longer doing research science or engineering work and was writing for journals, issuing lab publications, and reporting to clients. In her new position at US Customs and Border Protection, Novella is more of a senior adviser, advising on scientific matters and providing subject matter expertise for policy and strategy. Therefore, the labs are going to put her into a specialist rank. Her responsibilities include more than just managing money, as she liaises, plans strategy, provides technical assessments, and evaluates technologies.

Novella trains Customs and Boarder Protection officers and Border Patrol agents to use radiation portal monitors, handheld radiation scanners, and the pagers that check proximity to radiation. She was selected because of the need for double competency with the technology and education. They needed a teacher who was both theoretical and practical, able to not only give context and history, but also develop skills, and who was able to talk to anyone. (Novella credits her parents for the ability to make things simple and easy to understand.) The person who asked her to be on the team left and, eventually, she was asked to lead the team with its sizable budget, even though she was the youngest, and the only African American, on the team. After the 9/11 attacks, international customs groups heard about the training and wanted the technology. Among others, the Second Line Defense Program and Global Threat Reduction Initiative (GTRI) in DOE were interested. This increased the interest in training about radiation for agencies around the world. Novella's national lab and seven others formed teams to train international customs staff.

Novella does not mind working on many different projects. She feels one must learn the basic science and be able to apply it to many different things. But she did need to learn to budget her time. In the beginning, she would do one individual research project for two or more days and then go to another. As she began to get familiar with the projects, she could mix what she worked on in a day and still keep them straight. Later she found she could see the similarities between projects. Novella enjoys learning something new every day. She says, "that's what makes science fun." She's been able to work with radiochemists, nuclear engineers, systems engineers, physicists, and geologists, learning from each of them. She has enjoyed being in the national lab system because it is full of really brilliant people that are knowledgeable in their fields. She is grateful to be able to soak up everything they have to teach her.

The labs are now hiring younger scientists and getting into STEM. In general, they've found the pool of scientists is shrinking. They hire high school students as interns and, to keep them in the program, hire them again when they are undergrads in college. They have post-bachelor's programs for people with a bachelor's degree and post-master's programs for people with master's degree. Once the student is hooked, the labs try to keep them. Many students hired as post bachelors and post masters return to school for graduate degrees.

Novella has mentored both directly and indirectly. She has worked directly with the interns hired in her lab, helping them write research papers, create posters, and consider what to do about further schooling. Indirectly, having been in a similar situation, she speaks with groups of all the candidates the lab brings in as it seeks to diversify its workforce. Interacting with African Americans, Native Americans, and Latin Americans, during the summers she has been involved with a variety of activities. She meets personally with individuals, to share conversation and advice. She hosts barbecues and shares contact information. Many of them reach out to her by text or an instant message. Because she was a GEM Fellow, she always contacts GEM students. When Novella was first hired, she realized that the lab was promoting the fact that they had an African American scientist by taking her photo and advertising it. At first, she thought that maybe she should be reimbursed. But then she noticed that other minorities were also being tracked and realized that the students needed someone to look up to. Finally, she got all the minority scientists together and began meeting with the young interns and talking to them about a career in science and becoming a scientist. So, she took her own and others' experiences as minorities and made it positive for the students. She thinks that if the United States wants to be the "number one nation" in STEM, it will take all people, whatever their background, to do it, not just older white men in white coats.

Novella has been a member of the ACS since she was an undergrad. Jackson, while she was there, was trying to get a student affiliate group going, and the students regularly went to ACS meetings to present. In grad school, she really got active because her advisor, George Stanley, who was chair of the Baton Rouge Section and the Stanley Research group always volunteered for the ACS. After grad school she became a full member in the Richland Section but was picky about what she wanted to do. The average age of the Richland Section at the time she joined was sixty, and she really wanted to do promote learning projects for young scientists. This was a fairly large section, but Novella felt they were out of touch with what students needed. For National Chemistry Week, they would visit schools and give classroom talks. When she went with a group, she followed their talks with a demo which the students really liked. Afterward, Novella convinced the section to do demos and was asked to work on their education programs. Later she was asked to chair the section, becoming the first African American to hold the position.

She worked with the Women Chemists Committee, at first locally and then nationally, finding it very rewarding but frustrating. She thinks the ACS is stuck in a rut, not listening to the younger chemists coming into the field—people are doing non-traditional science and the ACS should not be so rigid in its thinking. In general, the public does not know the real meaning of science and what scientists do. However, she feels the millennial generation will change society as they think outside the box and push the envelope.

Novella started working with NSBE and NOBCChE as an undergrad because she was thinking about becoming a chemical engineer. In grad school, she was really active, serving as president of the NOBCChE chapter at LSU. After grad school, and in the west coast groups, she found the local organizations disconnected and disjointed, oriented more to students than professionals.

Novella is single, but not by choice. All the members of her family—parents, grandparents and siblings—married for life. She wants that too. When she moved to Washington State, there were few African Americans there which forced her to think about marrying outside her race and the impact of race in general. When she lived in Detroit there was a major race riot which scared her. She has seen a lot of animosity regarding race. She has been to private schools where she was one of only three or four African Americans in the school and been told derogatory things to her face by teachers, administrators, and others. In Washington State, she has dated outside her race but thinks the men dated her for the novelty, she being intriguingly different from the women they were used to. She does not want to date those who think of her as exotic or an oddity and so has focused on her career. When she has dated men of her own race the relationships were usually long distance from Seattle, WA, Portland, OR, or even Los Angeles, CA. After her move to Washington, DC, for her new job in September 2012, she was delighted by the available men, now that she is in her forties. She figures she still has time to have children because she was born when her mother was forty.

What has she contributed to the field? Basically, she said she thought she has made a beginning and wishes it were more. African American women always have to strive to be the very best, to the point that they always want to do more. She'd like to leave a lasting impact, because there are still so many minority children that don't have role models. She's confident society has gone past the time African American scientists were hired to fill quotas. Current minority scientists don't feel they need to prove themselves in the same way, by having to work harder than others to prove they are equally qualified.

She says the best way to change the numbers of minorities in STEM would be to catch students in elementary school beginning at the fourth to sixth grade. These students still want to learn, really absorb new information, and still find science intriguing. Action should specifically include speaking to girls that early, because somewhere in junior high they start thinking about boys and turn from math and science. They need to know it is more important to learn something than to worry about what they are wearing. But it's also time to break the stereotype of the unkempt woman scientist. After the early contact, it's important to continue working with the students all the way to college age.

She recommends two-year colleges for students who need remediation because it's less expensive. The opportunities there can strengthen skills and prepare for advancing to a full four-year college. This type of remediation would be unnecessary if programs in middle school and high school were strengthened. A high illiteracy problem also needs to be addressed.

As a member of the ACS WCC, which thinks about and works for the advancement of women, she'd like to see more concern about the education of children. Everyone in science, but especially women in science, should be working to ensure their children, grandchildren, nieces, and nephews study math and science and are encouraged to follow in the footsteps of the scientifically oriented.

She'd also like to see some changes for the new African American Museum of History in the District of Columbia. She recommends piquing kids' interest with features about less well-known black scientists, and their innovations. Some of the exhibits should have some of hands-on activities, so the kids could try what that person did. For example, if the exhibit included Dr. George Washington Carver, there should be something they could touch. Most people know that George Washing Carver work with peanuts but don't know that that work led to other inventions.

Novella is a member of First Baptist Church in the District of Columbia. The church was founded in the southwest DC by slaves and is about 152 years old. She works on the college prep ministry, striving to prepare high schoolers, especially seniors, for college. Adults are matched to teens to groom them in the summer before they become seniors in high school. They take them to college fairs, have work sessions, help students write their essays, and generally assist them with their applications. They bring in specialists from the Department of Education to help the students prepare their Federal Application for Federal Student Aid (FAFSA). The church also gives the students scholarships based on their level of participation in the ministry, giving points for each activity in which students participate, with the scholarship money based on the number of points the students have earned. The ministry also tries to help the students get scholarships from outside entities.

What advice would Novella give to someone looking to enter the field of chemistry? Tip number one: enjoy what you are doing. If you are someone that enjoys taking things apart, putting them together, and determining how things work, you are already a budding scientist. Tip number two: While you are still in middle school or high school look for summer science programs in your area and sign up for them. Explore where scientists can work. Understand that not all scientists work in labs, some are computer scientists, programmers, advisors, and so on. Tip number three: Get good at math because you are going to need it. Even if you don't like math you will eventually have to use math as a scientist. Also, don't characterize math and science, you might get the same kind of math in both your math and science classes. Tip number four: Keep working at math and science and choose the field that you are most interested in. Forget the monetary aspect of the field. Do something that you enjoy. This way going to work will be fun. When you enjoy what you do, it will always be your best job.[10]

Notes

1. GEM is a network of leading corporations, government laboratories, top universities, and top research institutions that enables qualified students from underrepresented communities to pursue graduate education in applied science and engineering.
2. http://patents.justia.com/patent/6537350
3. Dianne Gates Anderson, interview by Jeannette Brown, April 5, 2017, Oral History Transcript, Science History Institute, Philadelphia, PA.
4. Allison was born with a birth defect. She was born two months premature, which may be the cause of her disability. Her shoulder joints which are supposed to be ball and socket are fused so she has limited motion with her arms. She also has no depth perception and a problem with her eyes.
5. Chiarelli, Paul, Xiamen Gu, Allison A. Aldridge, and HuaPing Wu, "Matrix-Assisted Laser Desorption Ionization and Time-of-flight Mass Spectrometry for the Sensitive Determination of Acrylamide–Deoxynucleoside Adducts," *Analytica Chimica Acta* 368 (1998): 1–9.
6. Allison Ann Aldridge interview by Jeannette Brown, August 25, 2004 and August 23, 2009, Oral History Transcript, Science History Institute, Philadelphia, PA.
7. LaTonya Mitchell-Holmes interview by Jeannette Brown, March 25, 2015, Oral History Transcript, Science History Institute, Philadelphia, PA.

8. Lutheran Church–Missouri Synod (LCMS): The LCMS operates the largest Protestant school system in the United States. Currently the LCMS operates 1,368 early childhood centers, 1,018 elementary schools, 102 high schools, ten universities and two seminaries for a total of 2488 schools in the United States. These schools educate more than 280,000 students and are taught by almost 18,000 teachers.

9. Dr. Warner was a fundamental researcher who also worked on the mechanisms for maintaining and enhancing student education in STEM with a focus on minority students. Many have gone on to get PhDs.

10. Novella Bridges, interview by Jeannette Brown, February 27, 2015, Oral History Transcript, Science History Institute, Philadelphia, PA.

6

Life After Tenure Denial in Academia

6.1 Sibrina N. Collins

FIGURE 6.1. Sibrina N. Collins

6.1.1 Life After Tenure Denial

The year 2014 was absolutely devastating for me professionally and personally; I was denied tenure and I lost both my maternal and paternal grandmothers. Reflecting back on that time in my life, I am certain that I would not have been able to survive the experience without the support of my close family and friends. I truly believe that the story of my journey will help others experiencing difficult challenges in their careers.

6.1.2 Undergraduate and Graduate School Experiences

After graduating from Henry Ford High School in Detroit, MI, in 1988, I enrolled at Highland Park Community College (HPCC) in nearby Highland Park. My mother was working as a secretary in the nursing department at the time, so I was able to take advantage of the tuition benefit offered to the college's employees. I enrolled in a chemistry course for non-science majors, which I absolutely loved! Needless to say, after earning my associate's degree in 1990, I decided to pursue chemistry as a major. I enrolled at the University of Michigan-Dearborn and attended two semesters before transferring to Wayne State University (WSU), in Detroit.

My experiences as an undergraduate chemistry major at WSU led me on the path to pursue a doctorate in chemistry. In the fall of 1992, I was awarded an NIH-MARC (National Institutes of Health-Minority Access to Research Careers) Fellowship. This fellowship provided me not only funding support, but hands-on research training in the laboratory of Professor Regina Zibuck, a synthetic organic chemist. The environment in the Zibuck laboratory was very supportive and due to this mentoring experience, I wanted to earn a doctorate in chemistry. As a MARC Fellow, I was engaged in research and presented a poster on my research efforts at a national conference for the first time. Thus, I was developing fundamental laboratory and communication skills as an undergraduate researcher.

Also during this time at WSU, I became involved in the WSU-NOBCChE chapter, where I found a supportive network of African American students pursuing undergraduate degrees in chemistry. The chapter adviser was

Dr. Keith Williams, Director of Minority Student Initiatives in the chemistry department. In addition, Professor Joseph Francisco, past president of ACS, was a faculty member in the WSU department of chemistry during this time. Francisco, now Dean of the College of Arts and Sciences at the University of Nebraska, is the second African American to serve as president of ACS. The department of chemistry at WSU had a critical mass of African American students and African American faculty and staff. To this very day, Professor Francisco and Dr. Williams are important mentors to me. Thus, WSU provided a supportive environment for me to develop as a young chemist.

After graduating from WSU with a bachelor's degree in chemistry (*cum laude*) in 1994, I began graduate studies in the department of chemistry at Ohio State University (OSU), Columbus, OH.[1] At the time, I was interested in earning a doctoral degree in synthetic organic chemistry and my research mentor, Professor Regina Zibuck, encouraged me to apply to the program. OSU was the leading producer of African Americans earning doctoral degrees in the STEM disciplines for several years.[2] OSU graduates include Ruth Ella Moore (PhD, 1933), the first African American woman to earn a PhD in bacteriology, and chemists Robert Henry Lawrence Jr. (PhD, 1965), the first African American NASA astronaut, and Thomas Nelson Baker, Jr. (PhD, 1941), the first African American to earn a PhD in chemistry from the department of chemistry at OSU (Collins, 2011).[3] His father, Baker Sr., was the first African American to earn a doctorate in philosophy (from Yale University in 1903).

After arriving at OSU, I decided to pursue a doctorate in inorganic chemistry instead of organic chemistry, largely because of the beautiful colors of transition metal complexes. During my time at OSU, the African American students in the department established the first OSU-NOBCChE student chapter, and I proudly served as its first president. I earned my PhD under the direction of Professor Bruce Bursten, in March 2000, becoming the fourth African American woman to earn a doctoral degree in chemistry from OSU. My research focused on the low-temperature (96 K) matrix photochemistry of ruthenium dinuclear organometallic complexes (DOCs). The photochemical products generated were monitored experimentally using FT-IR (Fourier Transform-Infrared) spectroscopy. I am very proud of this research, which was later published in the *Journal of Cluster Science*.[4]

6.1.3 The Postdoctoral Years

After graduate studies at OSU, I accepted a postdoctoral appointment in the research laboratory of Dr. Isiah M. Warner, analytical chemistry professor and vice chancellor at LSU Office of Strategic Initiatives, for two reasons. First, LSU is one of the leading producers of minority PhD chemists[5] and second, Warner is a leading authority in the area of separation science. There, I received some good fundamental training on leading a laboratory and managing students.

The environment at LSU continued to fuel my passion for diversity issues, and I accepted an editorial position at the AAAS in Washington, DC, for a new web-based, journal called the *Minority Scientists Network* (*MiSciNet*, now *Science Careers*), which was published by *Science* magazine. The editorial content for MiSciNet included first-person testimonials and articles focused on diversity in STEM issues.[6] During my time at AAAS, Dr. Shirley Malcom, head of education and human resources there, also mentored me. Malcom is certainly a trailblazer in her own right, and I received some important lessons from her on being a woman of color in STEM.[7]

After completing my appointment at AAAS, I accepted my first teaching appointment at Claflin University, an HBCU in Orangeburg, SC. I always wanted to teach at an HBCU, and I was absolutely thrilled to join the department in 2003. It was exciting to mentor young people who reminded me of myself. In 2006, I left Claflin to pursue a position where I could focus solely on diversity in STEM. I served as Director of Graduate Diversity Recruiting at the University of Washington in Seattle, WA; my role primarily focused on enhancing relationships between STEM faculty there and faculty at Minority-Serving Institutions (MSIs). My experiences at Claflin served me well at UW. Moreover, we certainly had some key successes there, including establishing a chapter of SACNAS and departmental seminars from STEM faculty at Spelman and Morehouse Colleges.[8] As much as I enjoyed UW, I missed being in the classroom and working directly with students, and so I joined the Department of Chemistry at the College of Wooster, a small Primarily Undergraduate Institution (PUI) in northeast Ohio in August 2008.

6.1.4 Teaching and Mentoring at a PUI

There are many examples in the literature of the success of white male faculty mentoring underrepresented students in STEM fields.[9] For example,

Dr. Henry Gilman, an organic chemistry professor at Iowa State University recruited African American students to his laboratory in the 1930s;[10] one of Gilman's students, Samuel P. Massie Jr. (1946, PhD), was the first African American faculty member at the US Naval Academy.[11] Unfortunately, there are few examples in the literature of African American women describing their experiences training chemistry students.[12] I am proud to say that some of my former students have been successful obtaining entry-level positions in industry, while others completed doctoral degrees (inorganic chemistry) and degrees in law and the health sciences.

Mentoring students through undergraduate research experiences is essential to their success in chemistry.[13] My approach to training Wooster students in the laboratory was no different than my training of students at Claflin. Every single student is different and some students need more guidance than others in the laboratory. I have one goal, namely to help all of my students reach their full potential in the chemical sciences. I encourage my students by reminding them that conducting research is not easy and telling them about my own challenges as an undergraduate student. More importantly, I am proud to be a cheerleader for them, emphasizing important research milestones along the way. These are important lessons I have taken from my own mentors over the course of my career.

While teaching at Wooster, I established important research collaborations with faculty at Research One Institutions (R1 institutions), namely Dr. Claudia Turro, professor of chemistry at OSU and Dr. Bill Connick, professor in the department of chemistry at the University of Cincinnati. These collaborations have been fruitful leading to four peer-reviewed journal articles.[14] I also established a research collaboration with Professor Wiley Youngs at the University of Akron, and I am truly thankful to him for his support over the years.

Unfortunately, I was denied tenure in March 2014, but I am proud of my accomplishments during my time at Wooster. This was an extremely painful experience, but it forced me to stop running from who I am. I strongly believe that there is a lesson in every experience. Honestly, your happiness is your responsibility. For years, I was too afraid to admit what I really wanted to do for the rest of my life. I am not afraid anymore to walk in my truth.

From 2015 to 2016, I proudly served as Director of Education at Detroit's Charles H. Wright Museum of African American History, a leading cultural institution dedicated to the African American experience. As director of education, I focused on the science education and social studies pro-gramming for the K–12 community. I joined Lawrence Technological University (LTU) in Southfield, MI, as the first executive director of LTU's Marburger STEM Center on July 1, 2016, and I am enjoying this role.[15] I am responsible for promoting STEM[16] education on the campus, and I love this! This is exactly what I was born to do.

LTU's Marburger STEM Center is the intellectual home of the STEM and STEAM initiatives taking place on the campus. These initiatives in-clude our summer camps, which are taught by LTU faculty; Robofest, an international outreach program to teach young people how to program robots; and a new, innovative partnership with the Detroit Public Schools Community District (DPSCD), the Blue Devil Scholars Program (BDSP). The overall goal of BDSP is to enhance the STEAM curriculum in DPSCD schools by working effectively with teachers, students, and parents. This is an amazing opportunity to return to my hometown of Detroit and help make an impact on K–12 STEM education.

I am truly proud to say that I am a PhD chemist who is passionate about diversity in science and STEM education. Moreover, I have gained historical expertise for nearly two decades becoming a chemist-historian and publishing several articles focused on the important contributions of women and scientists of color. It is never easy to move forward from a dif-ficult career experience. However, I am walking testimony that you can be successful after a tenure denial.

Notes

1. S. N. Collins, "African Americans and Science," *Chemical & Engineering News* (October 26, 2009): 3.

2. (J.M. Jay, Negroes in Science: Natural Science Doctorates, 1876–1969 (Detroit, MI, Balamp Publishing, 1971).

3. S. N. Collins, "Celebrating Our Diversity: The Education of Some Pioneering African American Chemists in Ohio," *Bulletin for the History of Chemistry* 36, no. 2, (2011): 82–84.

4. S. N. Collins, C. M. Brett, and B. E. Bursten, "Density Functional Theory and Low-Temperature Investigations of CO-Loss Photochemistry from $[(C_5R_3)Ru(CO)_2]_2$ (R=H, Me) Complexes," *Journal of Cluster Science* 15, no. 4 (2004): 469–487.

5. S. N. Collins, G. Stanley, I. M. Warner, and S. F. Watkins, "What is Louisiana State University Doing Right?" *Chemical & Engineering News* (December 10, 2001): 39–42.

6. P. Limbach, "Can a White Male Really be an Effective Mentor?" *Science Careers* (2002). http://sciencecareers.sciencemag.org/career_magazine/previous_ issues/articles/2002_02_15/nodoi.14045371530930434946. Accessed June 2, 2013; C. Stewart, "The Unwritten Rules of Graduate School," *Science Careers* (2002). http://sciencecareers.sciencemag.org/career_magazine/previous_ issues/articles/2002_06_14/nodoi.12280453841658470941. Accessed June 2, 2013.

7. S. M. Malcom, P. Q. Hall, and J. W. Brown, *The Double Bind: The Price of Being a Minority Woman in Science* (Washington, DC: American Association for the Advancement of Science, 1976).

8. S. N. Collins, "Science Society's New Science Chapter Gaining Momentum," *UW Today* (May 24, 2007). http://www.washington.edu/news/2007/05/24/ science-societys-new-science-chapter-gaining-momentum-at-the-uw/. Accessed June 2, 2013).

9. C. L. Brown, "Henry Gilman," *Chemical & Engineering News* (May 14, 2001): 6; Limbach, "Can a White Male Really be an Effective Mentor?"

10. Brown, "Henry Gilman."

11. S. N. Collins, "The Gilman Pipeline: A Historical Perspective of African American PhD Chemists from Iowa State University," in *Chemistry at Iowa State: Some Historical Accounts of the Early Years* (Ames, IA: Iowa State University, 2006), 126–148.

12. J. E. Brown, *African American Women Chemists* (New York: Oxford University Press, 2012).

13. W. Pearson and I. M. Warner, "Mentoring Experiences of African-American PhD Chemists," in *Diversity in Higher Education*, ed. H.T. Frierson (Greenwich, CT: JAI Press Inc., 1998), 41–57; W. Pearson, *Beyond Small Numbers: Voices of African American PhD Chemists* (New York: Elsevier Press, 2005).

14. S. N. Collins, J. A. Krause, M. Regis, P. J. Ball, and W. B. Connick, "Self-assembly of a One-Dimensional Iron(II) Coordination Polymer with p-phenylenebis (picolinaldimine)," *Acta Crystallographica* C63 (2007): m528–m530; Collins, S.N., S. Taylor, J. A. Krause, and W. B. Connick. "2,6-bis(azaindole)pyri- dine: Reactivity with Iron(III) and Copper(II) Salts," *Acta Cryst. C. C63*, (2007): m436–m439; Y. Sun, S. N. Collins, L. E. Joyce, and C. Turro, "Unusual Photophysical Properties of a Ru(II) Complex Related to $[Ru(bpy)_2(dppz)]^{2+}$," *Inorganic Chemistry* 49 (2010): 4257–4262; J. A. Krause, D. Zhao, S. Chatterjee, R. Falcon, K. Stoltz, J. C. Warren, S. E. Wiswell, W. B. Connick, and S. N. Collins, "In-House and Synchrotron X-ray Diffraction studies of 2-Phenyl-1,10- Phenanthroline, Protonated Salts, Complexes with Gold(III) and Copper(II), and an Orthometallation Product with Palladium(II)," *Acta Crystallographica* C70 (2014): 260–266.

15. M. Lewis II, Lawrence Tech Names Marburger STEM Center's First Executive Director. *Crain's Detroit Business* (2016). http://www.crainsdetroit.com/article/20160422/NEWS/160429941/lawrence-tech-names-marburger-stem-centers-first-executive-director. Accessed April 22, 2016.

16. STEM means Science, Technology, Engineering, Mathematica/Science, Technology, Engineering, Mathematics.

7

Next Steps

THIS BOOK RELATES the stories of some amazing women who are currently working as chemists or are recently retired. These women, as I have said before, are hiding in plain sight. Perhaps the first or the only woman of color to work in a particular lab or university, they all managed to succeed in spite of any obstacles they faced. This chapter presents some ideas as to what you can do in order to succeed if you, your child, grandchild, or students might be interested in a STEM career, especially in chemistry.

I highly recommend reading Dr. Sandra L. Hanson's book, *Swimming against the Tide: African American Girls and Science Education*.[1] Dr. Hanson studied young African American girls in high school and their attitudes toward science, which has traditionally been a male profession. One of her conclusions is that these young girls need to see or read about role model, an African American woman chemist. *Swimming against the Tide* was written before the explosion of the World Wide Web and web-based materials, so this conclusion may no longer hold. For example, information about most of the women in this book is available on the web; some of them have given talks that are also on the web. Many of the women whose stories are told here work to mentor young minority students. These web based materials can be accessed by the students and teachers.

The primary organization that focuses on careers in chemistry is the ACS. What is chemistry? It is a branch of science that provides opportunities for a variety of careers, not just working in a research laboratory making new chemicals. According to the ACS, "In simplest terms, chemistry is the science of matter, for example anything that can be touched, tasted, smelled, seen or felt is made of chemicals."[2] The ACS website has information on chemistry careers, including videos of the different jobs that chemists do, and information on various technical disciplines,[3] as well as profiles of many chemists, including one whose life story is in this book.[4] As can be discovered by reading about the women in this book, chemists have a variety of careers. The ACS website states, "Chemists are the people who transform the everyday materials around us into amazing things.

In order to become a chemist, you have to want to learn how things work. Children are born to be scientists, from a young age always wanting to know what things are. Some of the women in this book have stated that they were curious about things in their early life. The ACS has books and programs for kids of all ages,[5] and the ACS education page has resources for students of all ages—elementary, middle, and high school, undergraduate, two-year college, and graduate education.[6] The ACS also has community outreach programs that promote chemistry to the general public, including Earth Week in the spring, National Chemistry Week in the fall, chemistry festivals and programs sponsored by local ACS sections, a Kids and Chemistry program that brings local chemists to work with kids in their communities, and Science Coaches who mentor local teachers who may teach any level of student elementary through high school.[7]

If you are an educator, or would like to become a middle or high school educator, there are workshops and grants for you to learn how to do so or to give you professional development credits—information about these programs is on the Chemistry Educators and Faculty page of the ACS website.[8] In 2014 the ACS founded the American Association of Chemistry Teachers (AACT),[9] which provides classroom resources, and professional development for teachers of elementary, middle, and high school and is the only one of its kind that does all that. If you are a teacher it would be worthwhile to join even if only for virtual professional development.

For high school students, chemistry clubs can be started in their schools. Project SEED[10] was "established in 1968 to help economically disadvantaged high school students expand their education and career outlook. The program provides opportunities for students who historically lack exposure to scientific careers to spend a summer conducting hands-on research with a scientist in academic, industry, or government research laboratories. Students receive a fellowship award for their efforts and a chance to receive a SEED college scholarship." A Project SEED scholarship is for high school senior students who have worked at least one summer at a science institute under the Project SEED program. Scholarships are restricted to students who plan to major in a chemical science or engineering field such as chemistry, chemical engineering, biochemistry, materials science, or some other closely related major. The scholarships are intended to assist former SEED participants in their transition from high school to college. Project SEED scholarships are nonrenewable and are only awarded to first-year college students. Selection is based on achievement in school, success in the Project SEED program, financial need, and intended chemical-related field of scientific study.[11]

The ACS also has programs for students when they enter college, including the ACS Scholars Program,[12] renewable scholarships for underrepresented minority students who want to enter chemistry or chemistry-related fields. Awards of up to $5,000 are given to qualified students: African American, Hispanic, or American Indian high school seniors or college freshman, sophomores, or juniors pursuing a college degree in the chemical sciences or chemical technology are eligible to apply.

There is also the ACS-Hach Land Grant Undergraduate Scholarships,[13] "awarded to undergraduate chemistry majors who attend one of [ACS'] seventy-two partner institutions and express an interest in teaching high school chemistry. Students are selected on the bases of chemistry aptitude, interest, and need. The selection process is administered by partner universities." The partner institutions are land grant colleges in every state in the union.

When students graduate to become chemists or chemical engineers, the ACS has many resources for them. The national Women Chemists Committee (WCC) focuses on all women in chemistry;[14] some local sections also have women chemists committees. The WCC has awards for young chemists as well, listed on their website. The ACS Committee on Minority Affairs (CMA) aims to lead change in institutional culture within the ACS and the chemical enterprise and achieve full participation and expression of intellectual and creative capacity of underrepresented minorities.[15] There is also a Younger Chemists committee[16] that focuses on the needs of all chemists under the age of thirty. The Women of Color Program[17] builds community, provides resources, and advocates for minority women chemists. The Chemists with Disabilities Committee envisions a time when all individuals, including those with disabilities, will advance the chemical enterprise by drawing on the full range of their talents.[18] One of the women in this book became and works as a chemist in spite of the fact she was born with a disability.

The ACS has a Diversity and Inclusion Task Force that works on making the ACS more diverse. One of the women in this book was the first chair of this group. The ACS Diversity Statement is: "The American Chemical Society believes that to remain the premier chemical organization that promotes innovation and advances the chemical sciences requires the empowerment of a diverse and inclusive community of highly skilled chemical professionals regardless of race, gender, age, religion, ethnicity, nationality, sexual orientation, gender expression, gender identity, presence of disabilities, educational background, and other factors. Chemical scientists rely on the American Chemical Society to promote inclusion and diversity in the discipline."[19] The ACS collects and distributes data about racial, ethnic, and gender disparities and works tirelessly to implement effective programs and initiatives to close gaps in the classroom and in the chemistry-related workforce. One organization focuses specifically on the needs of African American chemists, NOBCChE,[20] which came out of an Ad Hoc Committee for the Professional Advancement of Black Chemists and Chemical Engineers organized in April 1972 to establish an organization that would focus on the needs and careers of black chemists. That organization's first annual meeting was held in 1973 and it has been going strong ever since. The organization meets every year and awards grants to African American students. They have programs for middle and high school students. Recently they have reconnected with the ACS for joint programs. Their vision is to be an influential organization ensuring that African Americans and other people of color are fully engaged in shaping the global scientific community. Their mission is to build an eminent cadre of successful diverse global leaders in STEM and advance their professional endeavors by adding value to their academic, development, leadership, and philanthropic endeavors throughout the life-cycle of their careers. Each year during the National Conference the organization sponsors a "Science Fair and Science Bowl." The Science Fair provides a venue for the students to present original research through a poster competition in which students compete individually. The Science Bowl allows students to compete in teams of four against other teams in a round-robin academic quiz bowl. What makes this competition unique is that at least 20 percent of the questions are about African American inventors, scientists, and engineers. Both competitions are divided into junior (sixth–eighth grade) and senior (ninth–twelfth grade) divisions. Trophies, along with other prizes, are awarded to students in each level for both competitions. Students are honored at an awards luncheon, along with a science educational enrichment activity.

The US Department of Education currently sponsors Upward Bound grants that colleges can apply for to run programs for students. The Upward Bound Math and Science program is designed to strengthen the math and science skills of participating students, to help them recognize and develop their potential to excel in math and science, and to encourage them to pursue postsecondary degrees in math and science, and ultimately careers in the math and science profession.[21]

Iota Sigma Pi is a national honor society for women in chemistry. Its major objectives are to promote interest in chemistry among women students; to foster mutual advancement in academic, business, and social life; and to stimulate personal accomplishment in chemical fields. Iota Sigma Pi was founded in 1902 and was organized on a nationwide basis in 1916. It sponsors awards for chemists. The number of African American women in this organization is not known, but there are a number who are active members.[22]

For academic chemists, COACh was founded in 1998 to help women in academia by a group of senior women faculty in the chemical sciences from across the United States with a common concern about the gender-based obstacles that women scientists face in trying to attain their career goals. They sponsor workshops to help women get ahead in their careers.[23]

Association of Women in Science (AWIS) provides a unique national platform where essential research meets advocacy, innovation, and practice in supporting the advancement of women in STEM.[24]

For historians of either African Americans or of science or of women, or any combination of the above, there are resources as well. Many colleges have courses on those subjects. I have seen curriculum lists for courses from several colleges that are teaching the history of African Americans in science. There is a lot of information on the web about women in science now. A group of college libraries in Pennsylvania, Tri-College Libraries (Bryn Mar, Haverford, and Swarthmore), have produced an excellent research guide, People of Color in STEM,[25] which lists research books, organizations, and even books for kids.

Some public school districts also teach the history of African Americans in science. During African American History Month students may be given an assignment to find a woman in science. I have seen a lot of student projects on Dr. Marie Daly on the web. Some states, New Jersey being one, are writing curricula about African Americans in science and other fields to be taught in the schools.

As our young people move into the future, it is important for them to know where things stand as they begin—the environment for women scientists. In 1973, there was a symposium on women in science resulted in a report, issued in 1978, called *The Double Bind.*[26] In October 2009, the

Committee on Equal Opportunities in Science and Engineering (CEOSE) of the NSF decided to see if anything had changed since *The Double Bind* was published. CEOSE organized a conference called "Women of Color in Science, Technology, Engineering, and Mathematics (STEM)." The author attended this conference as a representative of the American Chemical Society. The women who came were not all African American, because the focus was on all women of color, but the information presented was not much different from that in the original report. Women were entering the field but were not advancing as quickly as their white colleagues, whether male or female. This realization led to the foundation of the ACS Women of Color Program discussed earlier.

In conclusion, as said by all of the women in this book, the world must believe that African American women are capable of becoming top-notch scientists. I hope that as readers of this book learn about the experiences of these women, they will become convinced that this is so. By overlooking the talents of African Americans, the United States may be wasting the potential of our young women. May equal access become routine.

Notes

1. Sandra L. Hanson, *Swimming against the Tide: African American Girls and Science Education* (Philadelphia: Temple University Press, 2009).
2. What Chemists Do, ACS web site, https://www.acs.org/content/acs/en/careers/college-to-career/video/what-chemists-do.html? Accessed June 1, 2017.
3. https://www.acs.org/content/acs/en/careers/college-to-career/areas-of-chemistry.html. Accessed June 3, 2017.
4. https://www.acs.org/content/acs/en/careers/college-to-career/chemists.html. Accessed June 3, 2017.
5. https://www.acs.org/content/acs/en/education/publications.htm. Accessed June 3, 2017.
6. https://www.acs.org/content/acs/en/education/resources.html. Accessed June 3, 2017.
7. https://www.acs.org/content/acs/en/education/outreach.html. Access June 3, 2017.
8. https://www.acs.org/content/acs/en/education/educators.html. Accessed June 3 2017.
9. https://teachchemistry.org/. Accessed June 3, 2017.
10. https://www.acs.org/content/acs/en/education/students/highschool/seed.html. Accessed June 3, 2017.

11. https://www.acs.org/content/acs/en/funding-and-awards/scholarships/projectseed.html. Accessed June 3, 2017.

12. https://www.acs.org/content/acs/en/funding-and-awards/scholarships/acsscholars.html. Accessed June 3, 2017.

13. https://www.acs.org/content/acs/en/funding-and-awards/scholarships/hachundergradscholarship.html. Accessed June 3, 2017.

14. http://www.womenchemists.sites.acs.org. Accessed June 3, 2017.

15. https://www.acs.org/content/acs/en/about/governance/committees/minority.html. Accessed June 3, 2017.

16. http://ycc.sites.acs.org/. Accessed June 3, 2017.

17. https://www.acs.org/content/acs/en/membership-and-networks/acs/welcoming/diversity/women-chemists-of-color.html. Accessed June 3, 2017.

18. https://www.acs.org/content/acs/en/about/governance/committees/cwd.html. Accessed June 3, 2017.

19. https://www.acs.org/content/acs/en/membership-and-networks/acs/welcoming/diversity.html. Accessed June 3, 2017.

20. http://www.nobcche.org/. Accessed June 3, 2017.

21. https://www2.ed.gov/programs/triomathsci/index.html. Accessed June 3, 2017.

22. http://www.iotasigmapi.info/. Accessed June 3, 2017.

23. https://coach.uoregon.edu/. Accessed June 3, 2017.

24. https://www.awis.org/. Accessed June 3, 2017.

25. http://guides.tricolib.brynmawr.edu/c.php?g=285559&p=1901689. Accessed June 17, 2017.

26. http://files.eric.ed.gov/fulltext/ED130851.pdf. Accessed June 3, 2017.

Selected Publications

The following publications are not meant to be a compressive list of the publications of each woman or even the publications of all the women in this book. In order to find more publications, you can google the name of the woman you are interested in or check the Sci Finder Database which is available in some libraries and especially in academic libraries.

ALLISON ANN ALDRIDGE

Aldridge, Allison A., Pieter Groenewoud, and Roger Halbert. "Ultra High Performance Liquid Chromatography in the Contract Manufacturing Environment." *Pharmaceutical Technology* 33, no. 3 (March 2, 2009): 3 pages. http://www.pharmtech.com/ultra-high-performance-liquid-chromatography-contract-manufacturing-environment.

Chiarelli, M. Paul, Xiaomei Gu, Allison A. Aldridge, and HuaPing Wu. "Matrix-assisted Laser Desorption Ionization and Time-of-flight Mass Spectrometry for the Sensitive Determination of Arylamide±deoxynucleoside Adducts." *Analytica Chimica Acta* 368, nos. 1–2 (July 17, 1998): 1–9.

AMANDA BRYANT-FREDRICH

Amato, Nicholas, and A. Bryant-Friedrich. "The Impact of Structure on Oxidatively Generated DNA Damage Products Resulting from the C3'-Thymidinyl Radical." *ChemBioChem* 14 (2013): 187–190. (Chosen as an editorial spotlight by the journal.)

Audat, Suaad A. S., CherylAnn Trzasko Love, Buthina A. S. Al-Oudat, and A. C. Bryant-Friedrich. "Synthesis of C3'-Modified Nucleosides for Selective Generation of the C3'-Deoxy-3'-Thymidinyl Radical: A Proposed Intermediate in LEE Induced DNA Damage." *Journal of Organic Chemistry* 77 (2012): 3829–3837.

Becker, D., A. Bryant-Friedrich, C. Trzasko, and M. Sevilla. "Electron Spin Resonance Study of DNA Irradiated with Argon Heavy Ion Beams: Evidence for Formation of Sugar/Phosphate Radicals." *Radiation Research* 160 (2003): 174. (32 citations)

Bryant-Friedrich, A. "Generation of a C-3'-Thymidinyl Radical in Single-Stranded Oligonucleotides under Anaerobic Conditions." *Organic Letters* 6 (2004): 2329. (10 citations)

Bryant-Friedrich, A., and R. Neidlein. "Synthesis and Chemical Reactions of New Ethynyl-Substituted 1,6-Methano[10]annulenes." *Synthesis* (1995): 1506.

Bryant-Friedrich, A., and R. Neidlein. "Syntheses and Properties of Donor/Acceptor Arylethynyl-Substituted 1,6-Methano[10]annulenes." *Helvetica Chimica Acta* 80 (1997): 1639.

Bryant-Friedrich, A., and R. Neidlein. "Syntheses and Reactions of Thio-substituted 1,6-Methano[10]annulenes." *Helvetica Chimica Acta* 80 (1997): 128.

Körner, S., A. Bryant-Friedrich, and B. Giese. "C-3'-a- and b-branched 2'-Deoxythymidines as Precursors for the Selective Generation of C-3'-Nucleoside Radicals." *Journal of Organic Chemistry* 64 (1999): 1559. (4 citations)

Lahoud, G., A. Hitt, and A. Bryant-Friedrich. "The Aerobic Fate of the C-3'-thymidinyl Radical in Single-stranded DNA." *Chemical Research in Toxicology* 19 (2006): 1630–1636. (5 citations)

Lahoud, G., J. Fancher, S. Grosu, B. Cavanaugh, and A. Bryant-Friedrich. "Automated Synthesis,Characterization and Structural Analysis of Oligonucleotide C-3'-Radical Precursors." *Bioorganic & Medicinal Chemistry* 14 (2006): 2581–2588. (5 citations)

al-Oudat, Buthina, Alex Salyer, Kevin Trabbic, and A. Bryant-Friedrich. "3'-Modified Oligodeoxyribonucleotides for the Study of 2-Deoxyribose Damage in DNA." *Bioorganic & Medicinal Chemistry Letters* 23 (2013): 854–859.

Shaik, Raziya, Matthew W. Ellis, Matthew J. Starr, Nicholas J. Amato, and Amanda C. Bryant-Friedrich. "Photochemical Generation of a C5'-Uridinyl Radical." *ChemBioChem* (2015). doi: 10.1002 /cbic.201500330. https://onlinelibrary.wiley.com/doi/pdf/10.1002/cbic.201500330

Zaidi, Rehana and A. Bryant-Friedrich. "The Effect of Reductant Levels on the Formation of Damage Lesions Derived from a 2-Deoxyribose Radical in ssDNA." *Radiation Research* 177 (2012): 565–572.

CHERLYNLAVAUGHN BRADLEY

Allred, A. Louis, Cherlynlavaughn Bradley, and Thomas H. Newman. "Attachment of permethylpolysilane groups to platinum by electroreduction of chloropermethyl-polysilanes. X-ray photoelectron spectroscopy of permethylpolysilanes chemically bound to electrode surfaces." *Journal of the American Chemical Society* 100 no. 16 (1978): 5081–5084.

Bradley, Cherlynlavaughn, and Douglas J. Schiller. "Determination of sulfur compound distribution in petroleum by gas chromatography with a flame photometric detector." *Analytical Chemistry* 58 no. 14 (1986): 3017–3021.

Bradley, Cherlynlavaughn, and Jon W. Carnahan. "Oxygen-selective microwave-induced plasma gas chromatography detector for petroleum-related samples." *Analytical Chemistry* 60 no. 9 (1988): 858–863.

Pauls, Richard E., Mark E. Bambacht, Cherlynlavaughn Bradley, Stuart E. Scheppele, and Donald C. Cronauer. "Distribution and characterization of phenolics in distillates derived from two-stage coal liquefaction, Energy Fuels." *Energy Fuels* 4 no. 3 (1990): 236–242.

DOROTHY PHILLIPS

Bouvier, E., P. Iraneta, U. Neue, P. McDonald, D. Phillips, M. Capparella, and Y.-F. Cheng. "Polymeric Reversed-phase SPE Sorbents—Characterization of a Hydrophilic-lipophilic Balanced SPE Sorbent." *LC/GC* 16, no. 5 (1998): S53–S58.

Bouvier, E., D. Martin, P. Iraneta, M. Capparella, Y.-F. Cheng, and D. Phillips. "A Novel Polymeric Reversed-phase Sorbent for Solid-Phase Extraction." *LC/GC* 15, no. 2 (1997): 152–158.

Bouvier, E., D. Martin, P. Iraneta, M. Capparella, Y.-F. Cheng, D. Phillips, and L. Bean. "A New Water-wettable Solid Phase Extraction Sorbent for the Analysis of Drugs in Biological Fluids." *Waters Column* 6, no. 3 (1996): 1–5.

Cheng, Y.-F., D. Phillips, and U. Neue. "Simple and Rugged SPE Method for the Determination of Tetracycline Antibiotics in Serum by HPLC Using a Volatile Mobile Phase." *Chromatographia* 44, no. 3/4 (1997): 187–190.

Cheng, Y.-F., D. Phillips, U. Neue, and L. Bean. "Solid-phase Extraction for the Determination of Tricyclic Antidepressants in Serum Using a Novel Polymeric Extraction Sorbent." *Journal of Liquid Chromatography & Related Technology* 20, no. 15 (1997): 2461–2673.

Cheng, Y.-F., D. Phillips, U. Neue, M. Capparella, and L. Bean. "Simple Eextraction Methods for the Determination of Drug in Serum." *American Biotechnology* 15, no. 13 (1997): 14.

Dion, D., K. O'Connor, D. Phillips, G. Vella, and W. Warren. "New Family of High Resolution Ion Exchangers for Protein and Nucleic Acid Purifications from Laboratory to Process Scales." *Journal of Chromatography* 535 A(1990): 127–145.

Neue, U., D. Phillips, T. Walter, M. Capparella, B. Alden, and R. Fisk. "Reversed-phase Column Quality and Its Effect on the Quality of a Pharmaceutical Analysis." *LC/GC* 12, no. 6 (1994): 468–480.

Phillips, D. "Performance of Symmetry® Narrow-bore Columns for Isocratic and Gradient Analyses." *Waters Column* 6, no. 1 (1996): 6–15.

Phillips, D., B. Bell-Alden, M. Cava, E. R. Grover, W. Mandeville, R. Mastico, W. Sawlivich, G. Vella, and A. Weston. "Purification of Proteins on an Epoxy-activated Support by High Performance Affinity Chromatography." *Journal of Chromatography* A 536, no. 1–2 (1991): 95–106.

Phillips, D., P. Cheli, D. Dion, H. Hodgdon, A. Pomfret, and B. San Souci. "Schemes for Efficient Protein Purification on a Family of Polymeric Ion Exchangers in glass Columns." *Journal of Chromatography A* 599 (1992): 239–253.

Swartz, M., M. Tomany, D. Phillips, and T. Tarvin. "Isolation and Purification Methodologies Using New Silica-based Ion Exchange Media." *BioChromatography* 2, no. 1 (1987): 38–45.

Young, M., D. Phillips, P. Iraneta, and J. Krol. "Mixed-mode Solid-phase Extraction and Cleanup Procedures for the Liquid Chromatographic Determination of Thiabendazole and Carbendazim in Fruit Juices." *Journal of AOAC International* 84 (2001): 556–561.

GILDA BARABINO

Barabino, G. A. "A Bright Future for Biomedical Engineering." *Annals of Biomedical Engineering* 41, no. 2 (2013): 221–222.

Barabino, G. A., C. Chien, and F. Yin. "Cellular and Molecular Engineering in Transition." *Cellular and Molecular Bioengineering* 6, no. 2 (2013): 117.

Bilgen, B., and G. A. Barabino. "Modeling of Bioreactor Hydrodynamic Environment and its Effects on Tissue Growth." *Methods in Molecular Biology* 868 (2012): 237–255.

Goldman, S. M., and G. A. Barabino. "Cultivation of Agarose-based Microfluidic Hydrogel Promotes the Development of Large, Full-thickness, Tissue-engineered Articular Cartilage Constructs." *Journal of Tissue Engineering and Regenerative Medicine* 11, no. 2 (2017): 572–581.

Goldman, S. M., and G. A. Barabino. "Hydrodynamic Loading in Concomitance with Exogenous Cytokine Stimulation Modulates Differentiation of Bovine Mesenchymal Stem Cells towards Osteochondral Lineages." *BMC Biotechnology* 16, no. 1 (2016): 10.

Goldman, S. M., and G. A. Barabino. "Spatial Engineering of Osteochondral Tissue Constructs through Microfluidically Directed Differentiation of Mesenchymal Stem Cells." *BioResearch* 5, no. 1 (2016): 109–117.

Green, M., I. Akinsami, A. Lin, S. Banton, S. Ghosh, B. Chen, M. Platt, I. Osunkwo, S. Ofori-Acquah, R. Guldberg, and G. Barabino. "Microarchitectural and Mechanical Characterization of the Sickle Bone." *Journal of the Mechanical Behavior of Biomedical Materials* 48 (2015): 220–228.

Perkins, K., K. Malone, and G. Barabino. "Missed Encounters: A Qualitative Study of Views of Faculty on Mentoring and Student Narratives on Race in Science Education." In *Managing Diversity in Today's Workplace*, edited by M. Paludi. Santa Barbara, CA: CLIO Inc., in press.

Yang, Y., and G. A. Barabino. "Differential Morphology and Homogeneity of Tissue-engineered Cartilage in Hydrodynamic Cultivation with Transient Exposure to Insulin-like Growth Factor-1 and Transforming Growth Factor-β1." *Tissue Engineering* Part A, 19, no. 21–22 (2013): 2349–2360.

Yang, Y., and G. A. Barabino. "Requirement for Serum in Medium Supplemented with Insulin-transferin-selenium for Hydrodynamic Cultivation of Engineered Cartilage." *Tissue Engineering* Part A, no. 17 (2011): 2012–2035.

Yang, Y., A. J. Lee, and G. A. Barabino. "Coculture-driven Mesenchymal Stem Cell-differentiated Articular Chondrocyte-like Cells Support Neocartilage Development." *Stem Cells Translational Medicine* 1 (2012): 843–854.

Yang, Y., M. B. Ard, J. T. Halper, and G. A. Barabino. "Type I Collagen-Based Fibrous Capsule Enhances Integration of Tissue Engineered Cartilage with Native Articular Cartilage." *Annals of Biomedical Engineering* 42, no. 4 (2014): 716–726.

MANDE HOLFORD

Anand, P., A. O'Neil, E. Lin, T. Douglas, and M. Holford. "Tailored Delivery of Analgesic Ziconotide across a Blood Brain Model Using Nanocontainers." *Nature Scientific Reports* 5 (2015): 12497. DOI:10.1038/srep12497.

Gorson, J., G. Ramrattan, A. Verdes, E. M. Wright, Y. Kantor, R. Srinivasan, R. Musunuri, D. Packer, G. Albano, W. G. Qiu, and M. Holford. "Molecular Diversity and Gene Evolution of the Venom Arsenal of Terebridae Predatory Marine Snails." *Genome Biology and Evolution* 7, no. 6 (2015): 1761–1778.

Holford, M., N. Puillandre, Y. Terryn, C. Cruaud, B. M. Olivera, and P. Bouchet. "Evolution of the Toxoglossa Venom Apparatus as Inferred by Molecular Phylogeny of the Terebridae." *Molecular Biology and Evolution* 26, no. 1 (2009): 15–25. DOI: 10.1093/molbev/msn211.

Holford, M., M. M. Zhang, K. H. Gowd, L. Azam, B. R. Green, M. Watkins, J. P. Ownby, D. Yoshikami, G. Bulaj, and B. M. Olivera. "Pruning Nature: Biodiversity-derived Discovery of Novel Sodium Channel Blocking Conotoxins from *Conus bullatus*." *Toxicon* 53, no. 1 (2009): 90–98.

Verdes, A., P. Anand, J. Gorson, S. Jannetti, P. Kelly, A. Leffler, D. Simpson, G. Ramrattan, and M. Holford. "From Mollusks to Medicine: A Venomics Approach for the Discovery and Characterization of Therapeutics from Terebrid Marine Snails." *Toxins* 8 (2016): 117.

NOVELLA BRIDGES

Aubry, D. N. Bridges, and G. G. Stanley. "Polar Phase Bimetallic Hydroformylation: Polar Phase Hydroformylation: The Dramatic Effect of Water on Mono- and Dirhodium Catalysts." *Journal of the American Chemical Society* 125, no. 37 (2003): 11180–11181.

De Jong, W. A., H. M. Cho, C. Z. Soderquist, N. N. Bridges, and J. Abrefah. "Experiments On, and Computational Modeling of NMR Parameters in Crystals Containing Lanthanides and Actinides." American Chemical Society Conference Proceedings, Washington, 2002.

De Jong, W. A., H. M. Cho, C. Z. Soderquist, N. N. Bridges, and J. Abrefah. "O-17 NMR in Uranyl Containing Crystals: Experiments and Computational Studies." American Chemical Society Conference Proceedings, Washington, 2003.

Fisher, D. R., A. Gutowska, and N. N. Bridges. "Novel Yttrium-90 Radionuclide Polymer Composites as Therapeutic Agents for Cancer Treatment." PNWD-SA-6461, Battelle—Pacific Northwest Division, Richland, Washington, 2004.

Gutowska, A., N. N. Bridges, K. R. Minard, B. Jeong, and Y. H. An. "Nanostructured Stimuli-Responsive Hydrogels for Biomedical Applications." Presented by A. Gutowska at the NIBIB-DOE Nanobiotechnology Workshop in Bethesda, Maryland on March 18, 2005. PNNL-SA-44447 .

Henry, C. A., and N. N. Bridges. "Synthesis Modification and Characterization for Evaluation of Nanoparticle Effects on the Human Respiratory System." Presented by C. A. Henry at the Mickey Leland Conference, in Houston, Texas on August 9, 2005. PNNL-SA-46530. (Also submitted as an article for the Annual Mickey Leland Conference proceedings.)

Herling, D. R., C. L. Aardahl, Y. Wang, G. D. Maupin, D. N. Tran, N. N. Bridges, J. E. Holladay, D. P. Mendoza, C. H. Peden, and J. Storey. "Reformer-assisted Catalysis for NOx Emissions Control." Presented by C. L. Aardahl at the Advanced Reaciprocating Engine Systems Conference in Diamond Bar, California on March 15, 2005. PNNL-SA-44818.

PATRICIA CARTER IVES SLUBY

Sluby, P. "Giles Beecher Jackson, Director-General of the Negro Development and Exposition Company." *Negro History Bulletin* 38, no. 8 (December 1975): 480–483.

Sluby, P. "James Jackson of Virginia." *National Genealogical Society Quarterly* 67, no. 2 (June 1979): 104–106.

Sluby, P. "Paths to the Past for Youth—It's in the Blood." *Bookmark* 49, no. 3 (Spring 1991): 183–188.

Sluby, P. "African American Brilliance." *Tar Heel Junior Historian* (Fall 2006). https://files.nc.gov/dncr-moh/African%20American%20Brilliance.pdf

Sluby, P. "Giles Beecher Jackson." In *African American National Biography*, Vol. 4, 440–442. Oxford: Oxford University Press, 2008.

Sluby, P. "The Entrepreneurial Spirit of African American Inventors." In *The Inventive Spirit of African Americans: Patented Ingenuity*, 1st ed. by P. Sluby. Wesport, CT: Preger Publishers, 2004.

SABRINA COLLINS

Collins, S. "Robert Percy Barnes: From Harvard to Howard University." *American Chemical Society, Bulletin for the History of Chemistry* 40 no. 1 (September 24, 2015): 37.

Collins, S. "Lesson Plans on Women and African Americans in the Physical Sciences." *Michigan Science Teachers Association Newsletter* (October 20, 2015): 19–20.

Collins, S. "Making an Impact with STEAM in the Community." *Michigan Science Teachers Association Newsletter* (October 20, 2015).

Collins, S. "Do We Learn Enough About African American Scientists?" *Diversity in Action* (January 1, 2016).

Collins, S. "Alice Augusta Ball: Chemical Drug Pioneer." *Undark Magazine* (May 12, 2016).

Collins, S. "Undark: Evelyn Boyd Granville." *Undark Magazine* (June 13, 2016).

Collins, S. "Unsung: James E. LuValle—Chemist and Olympic Athlete." *Undark Magazine* (August 15, 2016).

Collins, S. "Unsung: William Claytor." *Undark Magazine* (November 3, 2016).

Collins, S. "Unsung: Jewel Plummer Cobb." *Undark Magazine* (February 7, 2017).

Collins, S. "Unsung: Marie Maynard Daly." *Undark Magazine* (August 7, 2017).

Collins, S. "Diversity in the Scientific Community Volume 2: Perspectives and Exemplary Programs." In Critical Mass Takes Courage: Diversity in the Chemical Sciences, American Chemical Society Symposium Series, 165–177. Washington, DC: American Chemical Society, 2017.

SAUNDRA YANCY MCGUIRE

Mcguire, S. *Study Guide for Chemistry—An Introduction to General, Organic, and Biological Chemistry.* San Francisco: W. H. Freeman and Company, 1981.

Mcguire, S. "Hemisphericity and Chemical Education." Proceedings of the Alabama Center for Higher Education Annual Science Symposium, Montgomery, AL, 1985.

Mcguire, S. "Authoritarian Holistic Education: Efficient and Effective." In *Methods and Techniques of Holistic Education*, edited by I. L. Sonnier. Springfield, IL: Charles C. Thomas, Inc., 1985.

Mcguire, S. "Learning and Teaching Strategies." *American Scientist* 98, no. 5 (September–October 2010): 378–382.

Mcguire, S. "Teaching and Learning Strategies That Work." *Science* 325: 1203–1204.

Mcguire, S. "Using the Scientific Method to Improve Mentoring." *Learning Assistance Review, Journal of the National College Learning Center Association* 12, no. 2: 33–45.

Mcguire, S. Instructor's Guide for Chemistry: Principles, Patterns, and Applications edited by Bruce Averill and Patricia Eldredge. San Francisco, CA: Pearson Education Prentice Hall, 2006–2009.

Mcguire, S. Student Study Guide and Selected Solutions for Introductory Chemistry, 3rd ed. by Russo and Silver. San Francisco, CA: Benjamin/Cummings, 2000.

Mcguire, S. Complete Solutions Manual for Introductory Chemistry, 3rd ed. by Russo and Silver. San Francisco, CA: Benjamin/Cummings.

Mcguire, S. Instructors Teaching Guide for Introductory Chemistry, 3rd ed. by Russo and Silver. San Francisco, CA: Benjamin/Cummings.

Mcguire, S. "The Impact of Supplemental Instruction on Teaching Students How to Learn. New Directions for Teaching and Learning, Supplemental Instruction: New Visions for Empowering Student Learning, Marion Stone, Glen Jacobs, eds., 3–11 Jossey-Bass Wiley Periodicals, Inc, 2006.

Mcguire, S. *And Gladly Teach: A Resource Book for Chemists Considering Academic Careers* (contributing author). Washington, DC: American Chemical Society, 2006. https://www.acs.org/content/dam/acsorg/education/publications/and-gladly-teach-2nd-edition.pdf.

Mcguire, S. "Teaching Your Students HOW to Learn Chemistry." In *Survival Handbook for the New Chemistry Instructor*, edited by Diane M. Bunce and Cinzia M. Muzzi. Upper Saddle River, NJ: Pearson Prentice Hall.

Mcguire, S. "Teaching Students How to Learn Chemistry." *Strategies for Success Newsletter*. San Francisco, CA: Benjamin/Cummings.

SHARON NEAL

Neal, S. L., and B. A. Rowe. "Fluorescence Probe Study of Bicelle Structure vs. Temperature: Developing a Practical Bicelle Structure Model." *Langmuir* 19 (2003): 2039–2048.

Roach, C. A., and S. L. Neal. "Numerical Correction of Detector Channel Cross-Talk Using Full-Spectrum Fluorescence Correlation Spectroscopy." *Applied Spectroscopy* 64, no. 10 (2010): 1145–1153.

Rowe, B. A., C. A. Roach, J. Lin, V. Asiago, O. Dmitrenko, and S. L. Neal. "Spectral Heterogeneity of PRODAN Fluorescence in Isotropic Solvents Revealed by Multivariate Photokinetic Analysis." *Journal of Physical Chemistry A* 112, no. 51 (2008): 13402–13412.

Rowe, B. A., and S. L. Neal. "Photokinetic Analysis of PRODAN and LAURDAN in Large Unilamellar Vesicles Using Multivariate Frequency-Domain Fluorescence." *Journal of Physical Chemistry B* 110, no. 30 (2006): 15021–15028.

SHERRIE PIETRANICO-COLE

Gillespie, P., S. Pietranico-Cole, M. Myers, et al. "Discovery of Camphor-derived Pyrazolones as 11β-hydroxysteroid Dehydrogenase Type 1 Inhibitors." *Bioorganic & Medicinal Chemistry Letters* 24, no. 12 (2014): 2707.

Gillespie, P., R. A. Goodnow Jr., G. Saha, G. Bose, K. Moulik, C. Zwingelstein, M. Myers, K. Conde-Knape, S. Pietranico-Cole, and S. S. So. "Discovery of

Pyrazolo[3,4-d]pyrimidine Derivatives as GPR119 Agonists." *Bioorganic & Medicinal Chemistry Letters* 24, no. 3 (2014): 949.

Gordon, O., Z. He, D. Gilon, S. Gruener, S. Pietranico-Cole, et al. "A Transgenic Platform for Testing Drugs Intended for Reversal of Cardiac Remodeling Identifies a Novel 11βHSD1 Inhibitor Rescuing Hypertrophy Independently of Re-vascularization." *PLoS One* 9, no. 3 (2014): e92869.

Hirschmann, R. F., J. Hynes Jr., M. A. Cichy-Knight, R. D. van Rijn, P. A. Sprengeler, P. G. Spoors, W. C. Shakespeare, S. Pietranico-Cole, et al. "Modulation of Receptor and Receptor Subtype Affinities Using Diastereomeric and Enantiomeric Monosaccharide Scaffolds as a Means to Structural and Biological Diversity. A New Route to Ether Synthesis." *Journal of Medicinal Chemistry* 41 (1998): 1382.

Hirschmann, R. F., K. C. Nicolaou, S. Pietranico, et al. "Nonpeptidal Peptidomimetics with a β-D-Glucose Scaffolding. A Partial Somatostatin Agonist Bearing a Close Structural Relationship to a Potent, Selective Substance P Antagonist." *Journal of the American Chemical Society* 114 (1992): 9217.

Hirschmann, R. F., K. C. Nicolaou, S. Pietranico, et al. "De Novo Design and Synthesis of Somatostatin Non-Peptide Peptidomimetics Utilizing β-D-Glucose as a Novel Scaffolding." *Journal of the American Chemical Society* 115 (1993): 12550.

Kelly, M. J., S. Pietranico-Cole, J. D. Larigan, et al. "Discovery of 2-[3,5-dichloro-4-(5-isopropyl-6-oxo-1,6-dihydropyridazin-3-yloxy)phenyl]-3,5-dioxo-2,3,4,5-tetrahydro[1,2,4]triazine-6-carbonitrile (MGL-3196), a Highly Selective Thyroid Hormone Receptor β Agonist in Clinical Trials for the Treatment of Dyslipidemia." *Journal of Medicinal Chemistry* 57, no. 10 (2014): 3912.

Knapp, S., P. K. Kukkola, S. Sharma, and S. Pietranico. "N-Benzoylcarbamate Cyclizations." *Tetrahedron Letters* 45 (1987): 5399.

Lukacs, C. M., N. G. Oikonomakos, R. L. Crowther, L.-N. Hong, R. U. Kammlott, W. Levin, S. Li, C.-M. Liu, D. Lucas-McGady, S. Pietranico, and L. Reik. "The Crystal Structure of Human Muscle Glycogen Phosphorylase a with Bound Glucose and AMP: An Intermediate Conformation with T-state and R-state Features." *Proteins: Structure, Function, and Bioinformatics* 63, no. 4 (2006): 1123.

Martin, R. E., C. Bissantz, O. Gavelle, C. Kuratli, H. Dehmlow, H. G. Richter, U. Obst Sander, S. D. Erickson, K. Kim, S. L. Pietranico-Cole, et al. "2-Phenoxy-nicotinamides are Potent Agonists at the Bile Acid Receptor GPBAR1 (TGR5)." *ChemMedChem* 8, no. 4 (2013): 569.

Nicolaou, K. C., J. M. Salvino, K. Raynor, S. Pietranico, et al. "Design and Synthesis of a Peptidomimetic Employing betaβ-D-Glucose for Scaffolding." In *Peptides - Chemistry, Structure, and Biology, Proceedings of the Eleventh American Peptide Symposium*, edited by J. E. Rivier and G. R. Marshall, 881. Leiden: ESCOM, 1990.

Pietranico, S. L., et al. "C-8 Modifications of 3-alkyl-1,8-dibenzylxanthines as Inhibitors of Human Cytosolic Phosphoenolpyruvate Carboxykinase." *Bioorganic & Medicinal Chemistry Letters* 17, no. 14 (2007): 3835.

Tsao, G. A., G. A. Stahl, A. M. Lefer, J. S. Salvino, S. Pietranico, and K. C. Nicolaou. "Antagonistic Actions of Two New Analogs of PAF." *Journal of Lipid Mediators* 1 (1989): 189.

SONDRA BARBER AKINS

Akins, S. *Toluene Hygienic Guide*. Akron, OH: American Industrial Hygiene Association, 1984.

Akins, S. "Restructuring the Mathematics and Science Curriculum: Elementary Leadership Teachers' Perspectives." PhD dissertation Columbia University 1993.

Akins, S., comp. "William Paterson University Science Teacher Certification Program Folio and Rejoinders." Submitted to the National Council for Accreditation of Teacher Education (NCATE) and the National Science Teachers Association (NSTA), 2003–2004.

Akins, S. "Bringing School Science to College: Modeling Inquiry in the Elementary Methods Course." In *Exemplary Science: Best Practices in Professional Development*, edited by Robert Yager, 13–33. Virginia: National Science Teachers Association, 2005.

Akins, S. *Journeys in Learning and Teaching Science*. Xlibris Publishing, forthcoming.

LEYETE WINFIELD

Jackson, K., and L. Winfield. "Realigning the Crooked Room: Spelman Claims a Space for African American Women in STEM." *Peer Review* (2014): 16.

Mateeva, N., L. Winfield, and K. Redda. "The Chemistry and Pharmacology of Tetrahydropyridines." *Current Medicinal Chemistry* 12 (2005): 551–571.

Payton-Stewart, F., S. Tilghman, L. Williams, and L. Winfield. "Benzimidazoles Diminish ERE Transcriptional Activity and Cell Growth in Breast Cancer Cells." *Biochemical and Biophysical Research Communications* 450 (2014): 1358–1362.

Winfield, L. "Nucleophilic Aromatic Substitution, a Guided Inquiry Laboratory Experiment." *Chemical Educator* 15 (2010): 110–112.

Winfield, L., and F. Payton-Stewart. "Celecoxib and Bcl-2: Emerging Possibilities for Anticancer Drug Design." *Future Medicinal Chemistry* 4 (2012): 361–383.

Winfield, L., R. Gregory-Bass, J. Campbell, A. Watkins. "Characterizing Ligand Interactions in Wild-type and Mutated HIV-1 Proteases." *Journal of Computational Science Education* 5 (2014): 1–9.

Winfield, L., T. Inniss, and D. Smith. "Structure Activity Relationship of Anti-proliferative Agents Using Multiple Linear Regression." *Chemical Biology Drug Design* 74 (2009): 309–316.

Winfield, L., D. Smith, K. Halemano, and C. Leggett. "A Preliminary Assessment of the Structure–Activity Relationship of Benzimidazole-based Anti-proliferative Agents." *Letters Drug Design Discovery* 5 (2008): 369–376.

Winfield, L., C. Zhang, C. Reid, et al. "Synthesis and Dopamine Transporter Binding Affinity of 2,6-Dioxopiperazine Analogs of GBR 12909." *Medicinal Chemistry Research* 11 (2002): 102–115.

Winfield, L., C. Zhang, C. Reid, et al. "Synthesis, Lipo-philicity and Structure of 2,5-Disubstituted 1,3,5-dithiazine Derivatives." *Journal of Heterocyclic Chemistry* 40 (2003): 827–832.

Xu, L., S. Izenwasser, J. L. Katz, T. Kopajtic, C. Klein-Stevens, N. Zu, S. A. Lomenzo, L. Winfield, and M. L. Trudell. "Synthesis and Biological Evaluation of 2-substituted 3β-tolyltropane Derivatives at Dopamine, Serotonin, and Norepinephrine Transporters." *Journal of Medicinal Chemistry* 45 (2002): 1203–1210.

Yerokun, T., and L. Winfield. "LLW-3-6 and Celecoxib Impacts Growth in Prostate Cancer Cells and Sub-cellular Localization of COX-2." *Anticancer Research* 34 (2014): 4755–4760.

ETTA GRAVELY

Redd, Ginger, Thomas Redd, Tracie Lewis, and Etta Gravely. "Combining Educational Technologies for Student Engagement in the Chemistry Classroom." In *ACS Symposium Book—Technology and Assessment Strategies for Improving Student Learning in Chemistry*, edited by Madeleine Schultz, Siegbert Schmid, and Thomas Holme, 67–81. ACS Symposium Series 1235, 2016.

LATONYA MITCHELL-HOLMES

Gardner, J. Y., D. E Brillhart, M. M. Benjamin, L. G. Dixon, L. M. Mitchell, and J.-M. D. Dimandja. "The Use of GC×GC/TOF MS with Multivariate Analysis for the Characterization of Foodborne Pathogen Bacteria Profiles." *Journal of Separation Science* 34 (2011): 176–185. DOI:10.1002/jssc.201000612.

Bibliography

This is a bibliography for people who want to do further research on women chemists. This book was written from oral histories of the women profiled here. For people who want to listen to or read any complete oral history, those are archived at the Science History Institute and can be found at this website: https://www. sciencehistory.org/oral-history-collections, which also houses the oral histories of many women who are not profiled in this book.

AFRICAN AMERICAN WOMEN SCIENTISTS

Davis, Marianna W. *Contributions of Black Women to America.*" Vol. II. Columbia, SC: Kenday Press, 1982.

Gates, Henry Louis, and Evelyn Brooks Higginbotham. *The African American National Biography.* New York: Oxford University Press, 2008.

Jordan, Diann. *Sisters in Science: Conversations with Black Women Scientists about Race, Gender, and Their Passion for Science.* West Lafayette, IN: Purdue University Press, 2006.

Kessler, James H. *Distinguished African American Scientists of the 20th Century.* Phoenix, AZ: Oryx Press, 1996.

Krapp, Kristine. *Notable Black American Scientists.* Detroit, MI: Gale, 1999.

Lawrence-Lightfoot, Sara. "Toni Schiesler. My Mother's Power Was in Her Voice." In *I've Known Rivers: Lives of Loss and Liberation,* edited by Smith, Jessie Carney, and Phelps, Shirelle 197–223. Reading, Massachusetts: Addison-Wesley Publishing Company, 1994.

Sammons, Vivian Ovelton. *Blacks in Science and Medicine.* New York: Taylor & Francis, 1989.

Smith, Jessie Carney, and Shirelle Phelps, eds. *Notable Black American Women.* Detroit, MI: Gale, 1992.

Warren, Wini. *Black Women Scientists in the United States.* Bloomington: Indiana University Press, 1999.

Warren, Wini Mary Edwina. "Hearts and Minds: Black Women Scientists in the United States 1900–1960." PhD Diss., Department of the History and Philosophy of Science, Indiana University, 1997.

HISTORY

Manning, Kenneth R. "African Americans in Science." In *Ideology, Identity, and Assumptions*, edited by Howard Dodson and Colin A. Palmer, 49–95. New York: New York Public Library, 2007.

Manning, Kenneth R. "The Complexion of Science." *Technology Review* (November–December 1991): 61–69.

Pearson, Willie. *Beyond Small Numbers: Voices of African American PhD Chemists.* Amsterdam: Elsevier JAI, 2005.

Pearson, Willie. Black Scientists, *White Society, and Colorless Science: a Study of Universalism in American Science*. Millwood, NY: Associated Faculty Press, 1985.

Taylor, Julius H., and Morgan State College. *The Negro in Science*. Baltimore, MD: Morgan State College Press, 1955.

Young, Herman A., and Barbara H. Young. *Scientists in the Black Perspective.* Louisville, KY: Lincoln Foundation, 1974. https://www.loc.gov/rr/scitech/tracer-bullets/blacksinscitb.html.

WOMEN CHEMISTS

Ambros, Susan A., Kristin L. Dunkle, Barbara B. Kazarus, Indira Nair, and Deborah A. Harkus. *Journeys of Women in Science and Engineering*. Philadelphia: Temple University Press, 1997.

Bailey, Martha J. *American Women in Science: 1950 to the Present: A Biographical Dictionary*. Santa Barbara, CA: ABC-CLIO, 1998.

Science History Institute. *Her Lab in Your Life: Women in Chemistry.* Philadelphia: Science History Institute, 2005.

Fort, Deborah C, Stephanie J. Bird, and Catherine Jay Didion, eds. *A Hand Up: Women Mentoring Women in Science*. Washington, DC: Association for Women in Science, 1993.

Grinstein, Louise S., Rose K. Rose, and Miriam H. Rafailovich. *Women in Chemistry and Physics: A Bio-bibliographic Sourcebook*. Westport, CT: Greenwood Press, 1993.

Herman, Christine. *Iota Sigma Pi: National Honor Society of Women in Chemistry.* Radford, VA: Iota Sigma Pi, 2008.

Hinkle, Amber S., and Jody A. Kocsis. *Successful Women in Chemistry: Corporate America's Contribution to Science*. Washington, DC: American Chemical Society, 2005.

Moore, Patricia, Judith Love Cohen, and David A. Katz. *You Can Be a Woman Chemist.* Marina del Rey, CA: Cascade Pass, 2005.

Rayner-Canham, Marelene F., and Geoffrey William Rayner-Canham. *Women in Chemistry: Their Changing Roles from Alchemical Times to the Mid-twentieth Century*. Washington, DC: American Chemical Society, 1998.

EDUCATORS' RESOURCES

Advancing Diversity in the US Industrial Science and Engineering Workforce: Summary of a Workshop National Academy of Engineering National Academies Press, July 14, 2014. https://www.nap.edu/read/13512/chapter/1

Bayer Facts of Science Education, XIV, Female and Minority Chemists and Chemical Engineers Speak about Diversity and Underrepresentation in STEM. https://us2020.org/sites/default/files/Bayer_Facts_of_Science_Education_Executive_Summary.pdf.

Bernstein, Leonard, Alan Winkler, and Linda Zierdt-Warshaw. 1998. *African and African American Women of Science.* Maywood, NJ: Peoples Pub. Group, 1998.

Bystydzienski, Jill M., and Sharon R. Bird. *Removing Barriers: Women in Academic Science, Technology, Engineering, and Mathematics.* Bloomington: Indiana University Press, 2006.

Hanson, Sandra L. *Swimming against the Tide: African American Girls and Science Education.* Philadelphia: Temple University Press, 2009.

Hutson, Brittany. *Women of Power Adding Diversity into the Equation,* June 1, 2010. http://www.blackenterprise.com/adding-diversity-into-the-equation/.

Jemison, Mae. *Find Where the Wind Goes: Moments from my Life.* New York: Scholastic, 2001.

Kahn, Jetty. *Women in Chemistry Careers.* Mankato, MN: Capstone Books, 2000.

Malcom, Shirley, Paula Quick Ball, and Janet Walsh. *The Double Bind: The Price of Being a Minority Woman in Science.* Washington, DC: American Association for the Advancement of Science, 1976. http://www.eric.ed.gov/PDFS/ED130851.pdf. Accessed March 10, 2011.

Mosley, Pauline, and S. Keith Hargrove. *Navigating Academia: A Guide for Women and Minority STEM Faculty.* New York: Academic Press, 2014.

Sullivan, Otha Richard. *Black Stars: African American Women Scientists and Inventors.* New York: John Wiley & Sons, 2002.

Warmager, Paul, and Carl Heltzel. "Alice A. Augusta Ball Young. Chemist Gave Hope to Millions." *ChemMatters* (February 2007): 16.

Warren, Rebecca Lowe, and Mary H. Thompson. *The Scientist within You.* Eugene, OR: ACI Pub, 1995.

Warren, Rebecca Lowe, and Mary H. Thompson. *The Scientist within You, [vol. 2]: Women Scientists from Seven Continents: Biographies and Activities.* Eugene, OR: ACI Pub, 1996.

MEDIA

Archives of Women in Science and Engineering. *The Women in Chemistry Oral History Project*. Ames: Iowa State University, 2001. http://lib.iastate.edu/spcl/wise/Dreyfus/dreyfus.html.

McLean, Lois, and Richard Tessman. *Telling Our Stories: Women in Science*. Program Guide and CD. McLean Media, 1996, 1997. http://www.storyline.com.

The Science History Institute Oral History Project. *Women in Science*. http://chemheritage.org/research/policy-center/oral-history-program/projects/women-in-science.aspx.

The ScienceMakers. http://www.thehistorymakers.org/makers/sciencemakers.

Index